RULING DISTRIBUTED DYNAMIC WORLDS

RULING DISTRIBUTED DYNAMIC WORLDS

Peter S. Sapaty
Institute of Mathematical Machines and Systems
National Academy of Sciences of Ukraine

JOHN WILEY & SONS, INC.

This book is printed on acid-free paper. ∞

Copyright © 2005 by John Wiley & Sons, Inc. All rights reserved.

Published by John Wiley & Sons, Inc., Hoboken, New Jersey.
Published simultaneously in Canada.

No part of this publication may be reproduced, stored in a retrieval system, or transmitted in any form or by any means, electronic, mechanical, photocopying, recording, scanning, or otherwise, except as permitted under Section 107 or 108 of the 1976 United States Copyright Act, without either the prior written permission of the Publisher, or authorization through payment of the appropriate per-copy fee to the Copyright Clearance Center, Inc., 222 Rosewood Drive, Danvers, MA 01923, 978-750-8400, fax 978-646-8600, or on the web at www.copyright.com. Requests to the Publisher for permission should be addressed to the Permissions Department, John Wiley & Sons, Inc., 111 River Street, Hoboken, NJ 07030, (201) 748-6011, fax (201) 748-6008.

Limit of Liability/Disclaimer of Warranty: While the publisher and author have used their best efforts in preparing this book, they make no representations or warranties with respect to the accuracy or completeness of the contents of this book and specifically disclaim any implied warranties of merchantability or fitness for a particular purpose. No warranty may be created or extended by sales representatives or written sales materials. The advice and strategies contained herein may not be suitable for your situation. You should consult with a professional where appropriate. Neither the publisher nor author shall be liable for any loss of profit or any other commercial damages, including but not limited to special, incidental, consequential, or other damages.

For general information on our other products and services please contact our Customer Care Department within the U.S. at 877-762-2974, outside the U.S. at 317-572-3993 or fax 317-572-4002.

Wiley also publishes its books in a variety of electronic formats. Some content that appears in print, however, may not be available in electronic format.

Library of Congress Cataloging-in-Publication Data:

Sapaty, Peter.
  Ruling distributed dynamic worlds / Peter S. Sapaty.
    p. cm.
  Includes bibliographical references.
  ISBN 0-471-65575-9 (cloth)
  1. Electronic data processing--Distributed processing. 2. Mobile agents (Computer software) 3. Automatic control. I. Title.
  QA76.9.D5S28 2005
  004.3'6--dc22
                                              2004025909

Printed in the United States of America

10 9 8 7 6 5 4 3 2 1

To my grandson Eugene, the Angel

CONTENTS

Preface · xv

1 INTRODUCTION · 1

1.1 Toward Coordination and Management of Large Systems · 1
    1.1.1 Shifting from Computation to Coordination · 1
    1.1.2 Overoperability Versus Interoperability · 2
    1.1.3 Intelligent Systems Versus Intelligent Components · 3
    1.1.4 Directly Operating in Physical World · 4
    1.1.5 Distributed Artificial Life · 5
1.2 Problems of Managing Large Distributed Systems · 5
    1.2.1 From Localized to Distributed Solutions · 5
    1.2.2 More Distribution Problems and Details · 6
1.3 WAVE-WP: Basic Ideas · 8
    1.3.1 The Whole First · 8
    1.3.2 WAVE-WP Spatial Automaton · 9
    1.3.3 Implementation Basics · 9
1.4 Example: The Shortest Path Problem · 10
    1.4.1 Importance of Distributed and Parallel Solutions · 11
    1.4.2 Finding Shortest Path Tree · 11
    1.4.3 Collecting the Shortest Path Between Nodes · 13
    1.4.4 Main Problems of Distributed Implementation · 15
    1.4.5 Universal WAVE-WP Interpreters · 16
    1.4.6 Shortest Path Tree Finding in WAVE-WP · 17
    1.4.7 Shortest Path Collection in WAVE-WP · 19
    1.4.8 Full Program for Finding Shortest Path · 20
1.5 Example: Distributed Knowledge Representation and Processing · 21
    1.5.1 Knowledge Network · 21
    1.5.2 Elementary Query 1 · 22
    1.5.3 Elementary Query 2 · 24

1.6		System organization as a function of the application scenario	26
1.7		Relation to the Previous Book	26
1.8		Comparison with Other Works in Related Areas	27
	1.8.1	Parallel Computing	27
	1.8.2	Distributed Systems and Distributed Computing	27
	1.8.3	Parallel and Distributed Computing	28
	1.8.4	Computer Networking	28
	1.8.5	Intelligent Agents	28
	1.8.6	Mobile Agents	28
	1.8.7	Grid Computing	29
	1.8.8	Spatial Programming	29
	1.8.9	Mobile Robotics, Cooperative Robotics	29
	1.8.10	System Management	29
1.9		Organization of the Book	30

2 WORLDS AND WAVES IN THE WAVE-WP MODEL   33

2.1		Physical World	34
	2.1.1	Temporary Physical World Nodes	34
	2.1.2	Visiting Existing Nodes in a Region	35
	2.1.3	Destination Regions for New Nodes	36
	2.1.4	Accessing Physical World Parameters	36
	2.1.5	Broadcasting in Physical World	37
2.2		Virtual World	38
	2.2.1	Knowledge Networks	38
	2.2.2	Access to Nodes and Links	39
	2.2.3	Tunnel and Surface Broadcasting	40
	2.2.4	Linking with Alien Networks	41
2.3		United Physical–Virtual World	42
	2.3.1	The Integration Details	42
	2.3.2	Access to Nodes in the United World	43
	2.3.3	United World Dynamics	44
	2.3.4	Time and Speed	44
2.4		Execution World	44
	2.4.1	Doers and Their Connections	45
	2.4.2	Distribution of Physical–Virtual World Between Doers	46

		2.4.3	Absolute and Mapping Addresses	47
		2.4.4	Further Integration of Physical–Virtual–Execution World	47
	2.5	Waves		47
		2.5.1	Nature of Waves	47
		2.5.2	Navigation in Space	49
		2.5.3	Actions in Nodes	49
		2.5.4	Coverage with Rules	50
		2.5.5	Composition and Structuring of Waves	50
		2.5.6	Wave Expressions and Remote Data	53
		2.5.7	Delivery and Processing of Physical Matter	55
	2.6	Conclusions		56

3 WORLD PROCESSING LANGUAGE 59

	3.1	Top Language Organization		60
	3.2	Data Definitions		62
		3.2.1	General on Constants	62
		3.2.2	Special Constants	63
		3.2.3	Vectors	66
	3.3	Variables		66
		3.3.1	Nodal Variables	67
		3.3.2	Frontal Variables	67
		3.3.3	Environmental Variables	68
	3.4	Acts		73
		3.4.1	Flow Acts	74
		3.4.2	Fusion Acts	81
	3.5	Rules		85
		3.5.1	Rules in General	85
		3.5.2	State Generalization Procedure	85
	3.6	Forward Rules		86
		3.6.1	Branching Rules	86
		3.6.2	Repetition	89
		3.6.3	Synchronization	90
		3.6.4	Protecting Common Resources	90
		3.6.5	Network Creation	91
		3.6.6	Autonomy Rules	92
	3.7	Echo Rules		93

3.8	Expressions	95
3.9	Working with Physical Matter	96
3.10	Conclusions	97

4 DISTRIBUTED WAVE-WP INTERPRETATION IN DYNAMIC ENVIRONMENTS 99

4.1	Doers and Their Networks		99
4.2	Wave-WP Interpreter Architecture		101
	4.2.1	Main Interpreter Components	101
	4.2.2	Exemplary Interpretation Patterns	104
	4.2.3	Integration of the Interpreter with Other Systems	105
4.3	Track Infrastructure		106
	4.3.1	Forward and Backward Operations	106
	4.3.2	Optimization of the Track Network	107
4.4	Elementary Operations Involving Multiple Doers		109
	4.4.1	Local Operations in Doers	109
	4.4.2	Creating a New Virtual Node in Another Doer	111
	4.4.3	Moving into a New Physical Location	113
4.5	More Complex Spatial Operations		115
	4.5.1	Moving Data Through Tracks	116
	4.5.2	Migration of Knowledge Networks Between Doers	117
4.6	Other Distributed Interpretation Issues		118
	4.6.1	Mapping Strategies	118
	4.6.2	Dealing with Shortage of Vehicles	118
4.7	Conclusions		121

5 SPATIAL PROGRAMMING IN WAVE-WP 123

5.1	Traditional Sequential and Parallel Programming		123
	5.1.1	Programming in a Single Doer	123
	5.1.2	Programming in Multiple Doers	126
5.2	Virtual World Programming		130
	5.2.1	Creating Virtual World as a Knowledge Network	130
	5.2.2	Inhabiting the Virtual World with Mobile Entities	132
	5.2.3	Providing Openness of the Virtual World	134
	5.2.4	Observation of the Virtual World	135

	5.2.5	Distributed Inference in the Virtual World	137
	5.2.6	Mobility of the Virtual World in the Execution World	139
5.3	Mobility of Doers in Physical World		140
	5.3.1	Movement of a Single Doer	140
	5.3.2	Free Movement of Multiple Doers	141
	5.3.3	Synchronized Movement of Multiple Doers	141
	5.3.4	Movement of Multiple Doers by Turns	142
	5.3.5	Adding Payloads to Mobile Doers	143
5.4	Moving and Acting in Physical World Directly		144
	5.4.1	Sequential Movement in Physical World	145
	5.4.2	Parallel Movement in Physical World	147
	5.4.3	Combined Sequential–Parallel Movement	147
	5.4.4	Adding Payload: Planting Trees	150
5.5	Programming in Integration of Physical and Virtual Worlds		151
	5.5.1	Planting Trees in the United World	151
	5.5.2	Observation in the United World	153
	5.5.3	Programming of Spatial Dynamics	154
5.6	Conclusions		156

6 EXEMPLARY MISSION SCENARIOS 157

6.1	Coordinated Movement of a Group		158
	6.1.1	Stepwise Movement of a Two-Level Hierarchy	158
	6.1.2	Creation of a Persistent Infrastructure and Moving with It	159
	6.1.3	Extending to Any Number of Layers	160
	6.1.4	Simultaneous Movement of All Nodes in a Group	161
	6.1.5	Moving to the Averaged Positions of Subordinates	162
	6.1.6	Further Possible Group Movement Modifications	163
	6.1.7	Reverse or Heads-First Movement	163
	6.1.8	Movement in a Column	164
	6.1.9	Integrating Different Movement Solutions	166
6.2	Physical Matter Delivery and Remote Processing		167
	6.2.1	Most General Task Solution	167
	6.2.2	Splitting into Subtasks	167
	6.2.3	Adding Synchronization	168

		6.2.4	Setting Specific Routes	169
		6.2.5	Assigning Robots to Scenarios in Column Movement	169
	6.3	Physical World Search Assisted by Virtual World		171
		6.3.1	Creating the Distributed Virtual World	172
		6.3.2	Top-Level Space-Cleaning Scenarios	174
		6.3.3	Single-Step Multiple-Branch Search	175
		6.3.4	Full-Depth Search for Polygons	178
		6.3.5	Run Time Space Modification	179
	6.4	Map-Based Collection of Samples		181
	6.5	Conclusions		182

7 DISTRIBUTED MANAGEMENT USING DYNAMIC INFRASTRUCTURES 185

	7.1	Distributed Creation and Reconfiguration of an Infrastructure		186
		7.1.1	Hierarchical Infrastructure	186
		7.1.2	Other Topologies: Centralized, Chain, and Ring	187
		7.1.3	Infrastructure Modification	190
	7.2	Dynamic Hierarchy Based on Physical Neighborhood		192
		7.2.1	Finding the Most Central Unit	192
		7.2.2	Creating Infrastructure from the Center	193
	7.3	Basic Command-and-Control Scenario in WAVE-WP		195
		7.3.1	Recursive Hierarchical Command and Control (CC)	195
		7.3.2	Implementing CC in WAVE-WP	196
		7.3.3	Adding Payload to the CC Implementation	197
	7.4	Solving Distributed Management Problems		198
		7.4.1	Hierarchical Resource Management	198
		7.4.2	More Complex Management Scenarios	199
	7.5	Air Traffic Management in Dynamic Environments		201
		7.5.1	Creation of the Radar Neighborhood Infrastructure	201
		7.5.2	Mobile Tracking of an Aerial Object	202
		7.5.3	Simultaneous Multiple Tracking	203
		7.5.4	Setting Up Global Control	204
		7.5.5	Other Traffic Management Tasks	205
	7.6	Conclusions		206

8 MORE CRISIS MANAGEMENT SCENARIOS AND SYSTEMS — 207

- 8.1 Region Patrol by Mobile Robots — 208
  - 8.1.1 Patrolling by a Single Robot — 208
  - 8.1.2 Simultaneous Region Patrol by Two Robots — 209
  - 8.1.3 Possible Cooperation Between the Two Robots — 210
  - 8.1.4 Dynamic Patrol by Any Number of Robots — 211
- 8.2 Distributed Dynamic Cognitive Systems — 213
  - 8.2.1 Semantic Representation of Distributed Cognition — 213
  - 8.2.2 Multirobot Patrol as a Distributed Cognitive System — 214
- 8.3 Multirobot Hospital Scenarios — 215
  - 8.3.1 A Robotized Hospital — 215
  - 8.3.2 Hospital World Representation — 216
  - 8.3.3 State-Checking Scenario — 217
  - 8.3.4 Cleaning Scenario — 218
  - 8.3.5 Life Support Scenario — 219
  - 8.3.6 Multirobot Service Snapshot — 219
- 8.4 Future Combat Systems — 220
  - 8.4.1 Advantages of Using WAVE-WP — 220
  - 8.4.2 Target Fusion and Distribution by the Infrastructure — 221
  - 8.4.3 Fusion–Distribution Scenario in WAVE-WP — 222
- 8.5 Crises Management in Open Networks — 223
  - 8.5.1 Embedding Distributed WAVE-WP System — 223
  - 8.5.2 Establishing Higher Management Layer — 224
  - 8.5.3 Collecting All Infected Nodes — 225
  - 8.5.4 Finding Congested Links and Nodes — 226
  - 8.5.5 Inferring Possible Virus Sources — 227
  - 8.5.6 More Intelligent Solutions Required — 228
- 8.6 Using Global Infrastructures in WAVE-WP — 229
  - 8.6.1 Hypothetical Infrastructure Sketch — 229
  - 8.6.2 Air Defense Programming Example — 229
- 8.7 Conclusions — 231

9 CONCLUSIONS — 233

- 9.1 Summary of the Main Features of WAVE-WP — 233
  - 9.1.1 Starting from the Whole — 233
  - 9.1.2 The WAVE-WP Automaton — 234

	9.1.3	High-Level WAVE-WP Language	234
	9.1.4	Distributed WAVE-WP Interpreter	234
9.2	Some Main Application Areas		235
	9.2.1	Directly Accessing Physical World	235
	9.2.2	Distributed Knowledge Processing	235
	9.2.3	Operating in Physical World Under the Guidance of Virtual World	235
	9.2.4	Intelligent Network Management	236
	9.2.5	Advanced Crisis Reaction Forces	236
	9.2.6	Massive Cooperative Robotics	236
	9.2.7	Distributed Road and Air Traffic Management	237
	9.2.8	Autonomous Distributed Cognitive Systems	237
	9.2.9	Distributed Interactive Simulation	237
	9.2.10	Global Defense and Security	238
9.3	Final Remarks		238
	9.3.1	After the Final Remarks	239
9.4	Future Plans		239

APPENDIX: WAVE-WP SUMMARY 241

A.1	Extended Language Syntax	241
A.2	Compact Syntax Description	243
A.3	Permitted Abbreviations	243

References 245

Index 249

PREFACE AND ACKNOWLEDGMENTS

For the 5 years after *Mobile Processing in Distributed and Open Environments* was published, the interest in global networked solutions has grown enormously. Computers are now intensively acquiring bodies. As robots, they can also be mounted on humans or animals (even implanted). They can move and process not only information but physical matter or physical objects as well.

Crises relief, rapid reaction forces, future combat systems, and network-centric warfare are just a few hot areas where integral networked solutions are becoming of paramount importance. This encouraged us to attack much broader network-related problems with the spatial automaton called WAVE and to extend the model itself in order to cover with it any distributed worlds—physical, virtual, or combined, as described in the current book.

The extended model—actually a new one—WAVE-WP (or world processing) may treat the whole world (humans included) as a highly parallel and universal spatial machine that can be driven by a single spatial program. This is not a centralized approach, however, where the world is supposed to be coordinated from some central unit, say, a (symbolic) United Nations headquarters.

What WAVE-WP offers has no match to the existing human experience so far. It may be compared to global essence, or "soul," self-spreading and covering the whole world in the form of highly intelligent light, smog, smell, and if you want bad comparisons—radiation or a virus! This essence is fully distributed and parallel in nature and during the spread can modify the whole world or even create it from scratch. Many such spatial "souls," starting from arbitrary points of the distributed world, can evolve concurrently or competitively on the world's body, overlapping in time and space.

The technology described in this book allows us to express this global system essence on a topmost semantic level, automatically resulting in first generating and then harnessing this intelligent and ubiquitous smog or smell (or virus), while making it extremely useful, goal-oriented, and both internally and externally controllable. This can drastically improve the whole world and make it much better for everybody, the future generations particularly (that is why this book is dedicated to my grandson).

The book does offer a great power, which may happen to fall into both good and bad hands. The strength of the ideology and technology, in the author's opinion, may well exceed the power of nuclear energy if to extrapolate the latter onto the information field. By its self-spreading and penetrative power, WAVE-WP can gently cover and improve the world while orienting its evolution in a needed positive way, but it may also grasp and destroy the world's arbitrary parts or even the whole.

That is why continuing to keep hidden the key secrets of the full implementation (despite a number of trial, partial, realizations in different countries, with even a public domain version released) may be a right strategy, *in the sincere hope that the conditions will eventually arise in the better world for the full revelation and implementation.*

ACKNOWLEDGMENTS

In appreciation of many years of support of these ideas by the Institute of Mathematical Machines and Systems of the Ukrainian Academy of Sciences (as part of the V. M. Glushkov Center of Cybernetics), I thank its director Anatoliy Morozov and deputy Vitaliy Klimenko. The active cooperation with the Department of Intelligent Systems (with its rich history of creating intelligent hardwired computers, multiprocessor systems, and heterogeneous computer networks—the book inheriting these great projects), and numerous, often hot, discussions with its staff contributed much to the development of this book.

I also wish to thank Uliy Fisman, Victor Losev, Vladimir Voloboyev, Eugene Bondarenko, Vladimir Gritskov, Uri Kandzeba, Igor Gintov, Uri Korovitskiy, Nadezda Aronova, Anatoliy Belyaev, Ludmila Ostapchuk, and Vasiliy Kayurov. Cooperation from the *Journal of Mathematical Machines and Systems*—from Vitaliy Yashchenko and Raisa Dubina, international communications assisted by Galina Korotchenko and Irina Mitrofanova, as well as perfectly working Internet under Vitaliy Vishnevskey were crucial in moving this project forward.

Invaluable impact and inspiration, especially in the field of robotics and cooperative robotics, have been gained as the result of my one and a half year stay in Japan with Oita University. Personal thanks to Masanori Sugisaka who organized this cooperation, also Katsunari Shibata, Ms. Nagashima from VBL, as well as visiting Feng Xiongfeng from China. Multiple support of the book's topics by a number of Artificial Life and Robotics Symposia chaired by Prof. Sugisaka, and fruitful discussions with participants, as well as assistance from the AROB secretariat, contributed much to the robotized spirit of this book.

The stay at Oita University and the work on the book was with the support from the Ministry of Education, Culture, Sports, Science and Technology (Monbu-Kagakusho), as well as a grant from the Japan Society for the Promotion of Science (JSPS). Further influence on the book from artificial life topic, in integration with parallel and distributed processing, was by the started cooperation with the University of Aizu, where Nikolay, Mirenkov, and Minetada Osano provided particular moral support and practical help.

Participation in the world's largest Unmanned Systems Symposia and exhibitions in the United States, and communication with the AUVSI community assisted by Staci Butler, was important to understand and "touch by hands" the future of unmanned systems and mobile robotics, which inspired many robotic scenarios described in the book.

Years of direct or email communications with Robert Finkelstein from Robotic Technology, Inc., Kazuhiko Kawamura from Vanderbilt University, Jose

Delgado-Frias (Washington State University), Douglas Gage and John Bay (DARPA), Larry Deschaine (SAIC), Terence Lyons and Tae-Woo Park (AOARD), Kelley Myers (DMSO), John Blitch (moving with his rescue robots), Edward Feigenbaum (Stanford University), John Casti (Santa Fe), also Roland Vollmar from the University of Karlsruhe, Thomas Christaller from Fraunhofer Institute, and Werner Zorn from the University of Potsdam in Germany contributed much to the book's general orientation and final contents.

The recent ICINCO-2004 conference in Portugal and fruitful discussions with its Chair Joaquim Filipe, also Paula Miranda, and Filomena Pedroso from IBM, as well as frank exchange of opinions with Dinesh Kumar from Australia and Waseem Ahmed from Pakistan, influenced positively the final stages of the book's preparation.

The further development of the WAVE paradigm and its extended applications at UBC in Canada by Sergio Gonzalez Valenzuela and Son Vuong, and at Dublin City University in Ireland by Sanda Dragos and Martin Collier proved to be highly encouraging for writing a new book on this paradigm.

Everyday help from my wife Lilia and son Alexey, who (especially Lilia) chronically suffered from spoiled weekends and broken family plans and holidays (with myself barricaded in my study), was really heroic and invaluable.

PETER SAPATY

1

INTRODUCTION

We start with naming some exemplary areas that need radically new ideologies and technologies aimed at efficient organization of large distributed systems, explaining the existing problems of their understanding and management. Introductory ideas and simple practical examples explaining the essence of WAVE-WP (or World Processing), a novel technology that can rule a variety of distributed artificial and natural systems on a high, semantic level, are presented and discussed. Comparison with the predecessor model WAVE revealed in a previous book (Sapaty, 1999a), as well as the relation to other works in the area, is given, and the general organization of the book is outlined.

1.1 TOWARD COORDINATION AND MANAGEMENT OF LARGE SYSTEMS

1.1.1 Shifting from Computation to Coordination

The use of computers is steadily shifting from computations to coordination and management of complex distributed and dynamic systems, in both civil and military areas. Removing consequences of earthquakes and flooding, recovery after terrorist

---

Ruling Distributed Dynamic Worlds, by Peter S. Sapaty
ISBN 0-471-65575-9 Copyright © 2005 John Wiley & Sons, Inc.

attacks, or joint military operations in the world's hot spots are examples where distributed computerized systems may be particularly useful and efficient.

Both national and international campaigns of the scale and complexity never seen before may need to be planned, simulated, and managed, often within severe time constraints. These campaigns may use simultaneously within one system many human, technical, and natural resources distributed over large territories.

Radically new integration and coordination models and technologies may help such systems operate efficiently and fulfill their objectives, as traditional ones, stemming, say, from the Turing machine (Herken, 1995; Turing, 1936) or cellular automata (Wolfram, 1994) and oriented on computations, are becoming a real bottleneck.

1.1.2 Overoperability Versus Interoperability

Many current works in the field of integration and simulation of large systems orient on the support of interoperability between system components, where any unit can plug into a computer and see the same picture, say, of a battlefield. It is not easy to do, however, and billions of dollars are being spent on interoperability problems (Erwin, 2002).

This, however, may not be sufficient for the real integration, nor may be needed in general (Sapaty, 2002). For example, a tank driver should operate and make individual decisions within her direct vision only, and a local commander, executing orders from above, has to know the region where troops subordinate to her operate, rather than the whole battlefield.

The common picture is logically (not always practically, however) the simplest way to glue together unstructured or poorly organized systems. It reminds us, in some sense, the first multiprocessor computers which were designed decades ago. They were based on a common memory box, as the first feasible solution. In highly dynamic systems like real or symbolic battlefields, the huge amount of information they possess is an integral and inseparable part of the processes there. Making this information available for many players means that we must regularly create its snapshots and distribute them among the participants.

This may require highly intensive communications and broadband channels, which in severe and hostile environments may become a painful bottleneck. Also, as the information in such systems is changing with great speed, any snapshot of it may potentially be useless and "dead." And for security reasons, especially in military applications, the global system picture may need to be hidden rather than commonly available and shared.

Real integration may be based on more rational system principles, where parts are not necessarily equal to share the common picture (or do not share it at all, in the usual sense). They may rather form altogether a highly organized whole that defines the role of its parts and their interactions and not the other way round. The system's parts may lose their identity within such an organization, with their main goal being to satisfy the whole rather than evolve and prosper themselves.

Usually, such higher level organizations are a prerogative of human intelligence and manned command-and-control infrastructures. But emerging new types of distributed systems, especially those using automated and unmanned platforms and oriented on solving dynamic tasks in unpredictable environments, may require shifting of an essential part of this "overoperability" to the efficient automatic level.

Traditional agent-based philosophies and approaches (Earnshaw and Vince, 2002; Bigus et al., 2001; Lange and Oshima, 1998; Milojicic et al., 1999; Siegwart et al., 2004; Wooldridge, 2002), first designing parts and then assembling the whole from these parts in the hope it works properly, may not be able to cope with large dynamic systems, and higher level organizational models and technologies, directly supporting the system whole and guaranteeing the needed global behavior, are of growing importance.

1.1.3 Intelligent Systems Versus Intelligent Components

Theoretically, many jobs in dynamic environments can be performed by teams of robots better than by individual robots (Gage, 1993). And, technologically, any number of sophisticated mobile robots can be produced by industry today. But we are still far away from any suitable team solutions that could allow us to use this abundant and cheaper hardware efficiently, as the teamwork is a complex and insufficiently studied phenomenon, with neither universal nor even suitable results proposed so far.

A considerable increase in the robot's individual and group intelligence at the current stage may, however, be achieved by raising the level of language with which robots, especially their teams, are programmed and tasked. This can be used subsequently as a qualitatively new platform for the creation of advanced robotic systems that may integrate different existing control approaches or may be based on radically new, higher level organizational models.

We already have such precedents. Only with the introduction of high-level programming languages such as FORTRAN or LISP did real and massive use (as well as production) of computers began. It became possible to express the semantics of problems directly, abstracting from hardware details and drastically improving application programming productivity, with the system intelligence shifting from computer system components to programming languages and the quality of their implementation.

We can pursue a similar approach for multirobot systems, considering them not as a collection of intelligent individuals (which may never reach human- or even animal-like level), but rather as a parallel machine capable of executing any mission tasks written in a universal high-level scenario language (Sapaty, 2001a). Its automatic implementation, which may be environment dependent and emergent, can be effectively delegated to the distributed artificial brain embedded in, and formed by, multiple robots, as shown in Figure 1.1.

Having designed such a language, we now have the possibility of programming and organizing highly intelligent distributed systems as a whole, rather than assembling them from self-contained intelligent elements, which may be lost indiscriminately, say, on a battlefield, putting at risk the whole system. The needed system intelligence may always be kept *in the mission scenario* regardless of the run time composition of a

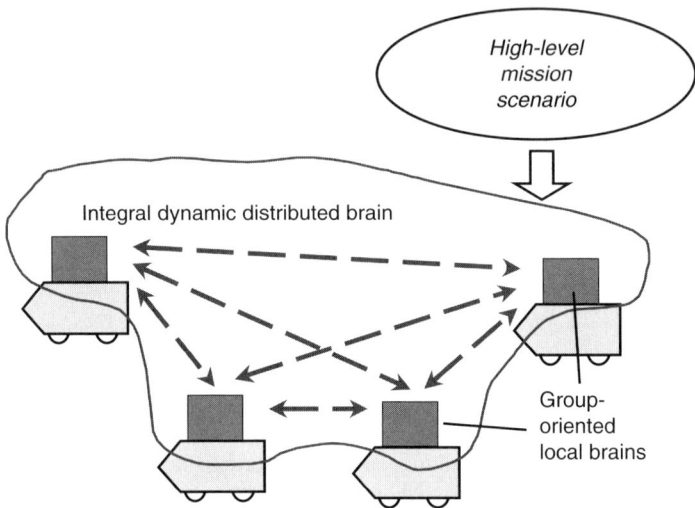

Figure 1.1. Group of robots as a parallel distributed machine.

system executing this scenario, thus enabling us to fulfill the mission objectives under any circumstances.

1.1.4 Directly Operating in Physical World

We also need models that can describe much broader activities than traditional information processing, enabling us to operate in real worlds directly, with physical movement and physical matter and object transference and manipulation.

Let us consider some examples of doing jobs in a physical world that may hint at how world-processing models and languages can be organized. For instance, if we want to perform a job, we first find a proper *place*, say, by a landmark or deviation from the current place (a hill at the right, 600 meters northwest, etc.), then move there with *operations* (machinery or equipment), and *data*, which may comprise both information and physical matter (a written plan of the job and, say, sand, bricks, water, ammunition, etc.).

Having performed the job (building a house, firing ammunition) using the data brought and/or existing there (say, left by other works), we may leave new data and the changes in the environment (the house built), pick up other data (unused materials), and move to other points, together with other operations (fresh or released equipment), to do another job (building a new house or a bridge), and so on. We may also need to bring results back to some point to be analyzed and processed (sand or water collected remotely, weapons or information seized in a reconnaissance operation, soil samples returned by a rover to a lander) to do other jobs or move further in case of success or failure.

Many jobs can be done simultaneously in the same or in different locations in the physical world. In performing jobs, we may cooperate or compete with others, seeing where, what, and how they do (moving in a chain, e.g., needs leveling and keeping distance with neighbors). Moving and doing jobs may depend on results and states reached in other places (deployment of a combat group only after success of another group, or entire mission abortion if the headquarters has been destroyed). Global control over evolving jobs may coexist with autonomous decision making.

To make real progress in this direction, we may need the design of a universal model, language and technology that would allow for an efficient description and implementation of a variety of parallel activities in a distributed physical world, similar to those sketched above.

1.1.5 Distributed Artificial Life

Life on Earth is a collective rather than individual phenomenon. It is also a distributed one. Different biological species can survive only if they communicate, and there are enough of them, being efficiently spread over vast territories. Artificial life approach (Langton, 1997; Ward, 2000) is so far considered mostly the creation and reproduction of individual life and single robotic species.

It is the right time to investigate a higher level now where distributed artificial systems and multiple robots can be really useful and can evolve if applied massively and work cooperatively. New advanced approaches for describing distributed autonomous systems on much higher levels than usual may be needed.

This may allow us to comprehend the very sense and basic evolution mechanisms of artificial life on a semantic level and also find, for instance, the *critical robotic mass* for this life to exist and prosper, using for simulation both single and parallel computers as well as computer networks.

The areas listed above are only a few of many where the design, or even invention, of radically new system creation and management models and technologies, like the one described in this book, are of highest interest and importance at present.

1.2 PROBLEMS OF MANAGING LARGE DISTRIBUTED SYSTEMS

Single-machine solutions often exhibit highest possible integrity as a system, with all resources at hand, and control being global, direct and absolute. On the contrary, the creation and management of large distributed systems that should behave as a single controllable entity pursuing global goals and recovering from damages can meet serious difficulties. Let us explain why.

1.2.1 From Localized to Distributed Solutions

In Figure 1.2 a symbolic splitting of the single-machine functionality into pieces for their distribution is shown, which may be based on separating by operations, separating by data, or by both.

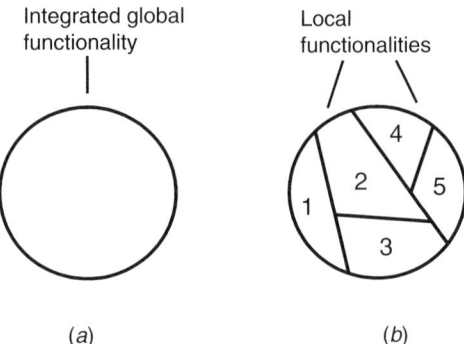

Figure 1.2. Partitioning of the problem for distribution: (a) single-machine solution and (b) breaking into pieces.

The resultant local functionalities may need to exist in more than a single copy in the distributed system (i.e., they should be replicated and spread throughout the space rather than localized and shared) for performance optimization, say, to become closer and work directly with other parts, and also to operate simultaneously with each other.

To work together as a distributed system, the parts of Figure 1.2b may need scores of other things to be additionally set up and integrated with them. First of all, there should be *communication and synchronization*, allowing them to exchange data via the *data channels*, to be established too. Second, the parts should have between (and over) them a sort of *command-and-control (CC) infrastructure*, to provide both local and global coordination and the ability to pursue common goals and obey global instructions. Within this infrastructure, additional *control centers* may be needed along with special *control channels* between them, and the control network may be multilayered for complex systems.

We have mentioned only a few additional things to be set up over the basic, split into pieces, functionality to work as a system, but even these can make the distributed solution very complex, as shown symbolically in Figure 1.3. Many other organizational tools and their interactions may be needed for this distributed system to operate properly (like, e.g., a recovery system, which may be distributed too and deeply integrated with both local functionalities and the distributed control).

1.2.2 More Distribution Problems and Details

Distributed Algorithms May Differ. The other difficulties may be in that the distributed algorithms may differ essentially from the centralized sequential ones (in our case of Fig. 1.2a), and they may need updating if not full rewriting of the functional parts shown in Figure 1.2b (as well as, possibly, introduction of other parts) to operate in a distributed and parallel manner. To bring the organization of Figure 1.3 to life, we may need to use together quite different existing languages and

MANAGING LARGE DISTRIBUTED SYSTEMS

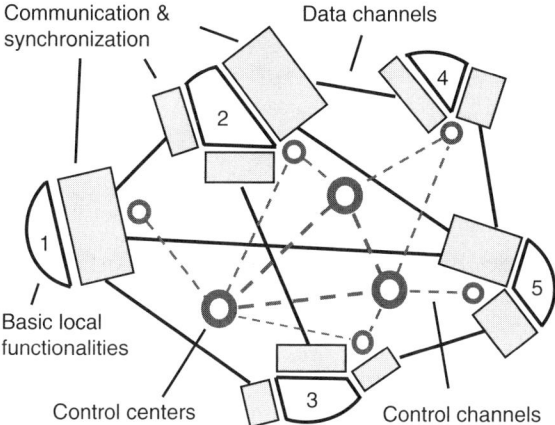

Figure 1.3. Overhead of distributed system solutions.

tools, and the whole system may become hugely overwhelmed with seams, scars, patches, and multiple interfaces.

Working in the Real World. In the real world, these functional pieces may have to acquire bodies and move physically, access and process not only information but physical matter or physical objects as well, also exchange the matter with each other along with the information, and so on. Additional control nodes and their communications with the original functional units and between them may require special (possibly, mobile) hardware units dedicated to command and control in the physical word.

Overwhelming Overhead. Creation, understanding, debugging, and subsequent managing of such distributed and dynamic systems may become a very complex problem, with the complexity growing drastically with the scale of problems to be solved. The organizational overhead, shown in Figure1.3, may considerably overwhelm the useful functionality of integral localized solutions, as in Figure 1.2*a*.

Starting from Scratch for Each Problem. And for each new problem (at least for each class of them), the whole work of creating, assembling, and debugging the complex distributed organization of Figure 1.3 must be repeated, possibly, from scratch. This is because *the local functions and their interactions, also the needed control over them, may be unique for each problem to be solved.*

Other Philosophies Needed. For solving complex problems in real, especially unpredictable and hostile, environments, we may need distributed systems organized in a very different manner than described above, with much higher clarity, flexibility, and also capability to recover after damages. This may require inventing, prototyping, and testing of quite different organizational ideologies, methodologies, and technologies than usual, as well as a possible revision of the ruling philosophies of existence, evolution, and general behavior of large dynamic and open systems, both manned and unmanned.

1.3 WAVE-WP: BASIC IDEAS

WAVE-WP is a completely different paradigm than any existing approaches oriented on distributed computation, coordination, and control. We provide here a sketch of some of its basic ideas and features.

1.3.1 The Whole First

Any existing approaches to investigation and management of large systems are *analytical* in nature, that is, first breaking the system into self-contained pieces or *agents* (or creating them from the start), studying them in detail, and then trying to assemble from these pieces the system as a whole, *seeing this whole from the level of these pieces*.

However, when the number of agents and interactions between them are getting large enough and these vary at system run time, serious problems of providing and supporting the needed global behavior arise, with the system as a whole becoming clumsy, unpredictable, and obscured by numerous details and routines (as already shown in the previous section).

The dominant parts-to-whole approach is becoming especially inefficient when these systems must operate in rapidly changing and hostile environments, where different parts (agents) can be destroyed, with the necessity of the whole to survive and fulfill the mission objectives.

The WAVE-WP (or World Processing) model attacks the problem by starting from the opposite side: *from the whole*. It allows us to grasp and express it on a *semantic level* in a high-level spatial language, abstracting from possible parts and their interactions (which appear and make sense in the model only in proper time and places), delegating the lower system organization to an efficient automatic implementation.

This, however, is *not a centralized approach*, where this whole as some global program or schedule is associated with a certain control center driving the distributed system, but rather is the *higher level, essence* of the system organization and behavior. This essence, which we symbolically call a *spatial gestalt*, is expressed in a special parallel and distributed formalism, which is quite different from conventional terms. [The term *gestalt* usually stands for a *form, shape, pattern*, as well as *organized whole*. Gestalt psychology (Davidson et al., 1989), see also the previous book (Sapaty, 1999a), represented a revolt from the atomistic outlook on the systems, starting with the organized whole as something more than the sum of the parts into which it can be logically analyzed.]

WAVE-WP *converts the whole distributed world into a universal spatial supercomputer*, which may operate with both information and physical matter and can include humans as elementary "processors" as well. The paradigm often enables us to program this world with the clarity and easiness we program tasks in single computers, using traditional programming languages.

1.3.2 WAVE-WP Spatial Automaton

The WAVE-WP automaton effectively inherits the integrity of traditional sequential programming over localized memory, but for working now with the real distributed world, while allowing its parallel navigation in an active pattern flow and matching mode—as a single and integral spatial process.

The automaton may start from any point of the distributed world or system to be controlled, dynamically covering its parts or the whole, and mounting a variety of parallel and distributed knowledge and control infrastructures. Implanting distributed "soul" into the system organization, the automaton may increase the system's integrity, capability of pursuing local and global goals, assessing distributed situations, making autonomous decisions, and recovering from indiscriminate damages.

In quite different applications, the automaton may need to destroy the very system it propagates through and/or operates in (or certain, e.g., malicious incursions or infrastructures in it). Many spatially cooperating or competing parallel WAVE-WP automata may evolve on the same system's body serving, say, as deliberative, reactive, and/or reflective spatial processes.

1.3.3 Implementation Basics

On the implementation layer, the WAVE-WP automaton widely uses high-level mobile cooperative program code that is self-spreading and replicating in networks and can be easily implemented on any existing software or hardware platform. As the automaton can describe direct movement and processing in the physical world, its implementation may need to involve multiple mobile hardware—with or without human participation.

A network of communicating (hardware or software) WAVE-WP language interpreters (or WI), which can be mobile if installed in manned or unmanned vehicles, should be embedded into the distributed world to be controlled, placing WIs into its most sensitive points, as shown in Figure 1.4.

During the spatial execution of system scenarios in WAVE-WP, individual interpreters can make local information and physical matter processing as well as physical movement in space. They can also partition, modify, and replicate program code, sending it to other interpreters (along with local transitional data), dynamically forming track-based distributed interpretation infrastructures.

The automaton can also exploit other systems as computational and control resources, with or without preliminary consent, that is, in a (controlled) viruslike mode. For example, existing network attacks (especially Distributed Denial of Service, or DDoS) may be considered as a possible malicious (simplified, degenerated, and distorted) implementation of the automaton.

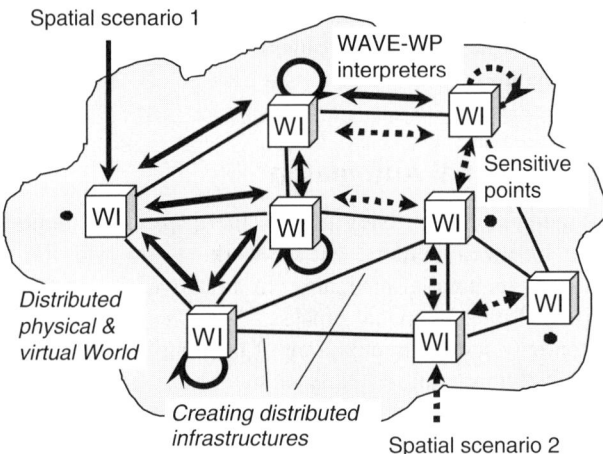

Figure 1.4. Network of interpreters with self-spreading spatial scenarios in WAVE-WP.

To explain the key WAVE-WP ideas further, let us consider the detailed expression in it of some well-known problems for which parallel and distributed solutions are of particular significance.

1.4 EXAMPLE: THE SHORTEST PATH PROBLEM

The first example relates to the finding of a shortest path between certain nodes in a network of weighed links (an example of which is shown in Fig. 1.5). This problem had been described in the previous book in relation to WAVE (Sapaty, 1999a), but due to its high practical importance and convenience in explaining the very essence of the paradigm, it is repeated in an extended form and with more details for WAVE-WP too, and within a broader context than before.

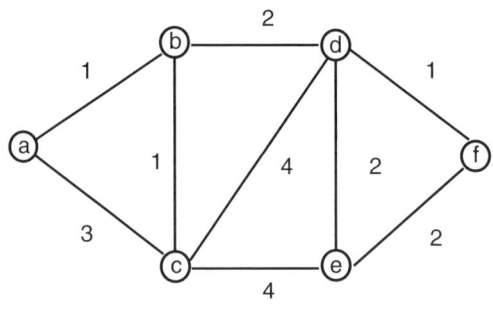

Link weights

Figure 1.5. Network with weighed links.

EXAMPLE: THE SHORTEST PATH PROBLEM

What is shown in Figure 1.5 may be a computer network in which the link weights may define the time of sending of a standard information packet between adjacent nodes, or a road network with the weights defining distance (say, in kilometers) between the road junctions, represented by nodes. Many other important optimization problems can also be converted into finding shorted paths within different time and space continuums or their combinations.

1.4.1 Importance of Distributed and Parallel Solutions

Distributed solutions of the shortest paths may be highly important in large dynamic spaces, where communications between computers can vary at run time (caused, say, by mobility of nodes, line congestions, or as a result of virus attacks). This may also relate to road networks (with embedded local computers serving parts of them), where roads or their junctions may get damaged in natural or human-caused disasters.

Collecting all needed information in one point and then solving the shortest path problem there may not be possible, even in principle, as the network topology and weights of links may be rapidly changing over time. So the solutions must be found directly where the information resides, as the latter is inseparable from the world it represents, losing its worth when accessed remotely. And parallel solution, in addition to the distributed one, may speed up the obtaining of the result considerably.

Let us find a shortest path (which may be more than one, in general) between nodes a and e of the network of Figure 1.5.

1.4.2 Finding Shortest Path Tree

We begin with finding a shortest path tree (or SPT) starting from node a and covering the whole network. This tree will directly show the shortest paths from the starting node to all other nodes in the network (node e included), and every such path can be easily traced by just moving from each of these nodes up the tree to its root.

The most natural expression of the general solution of this problem may be as follows. By starting in node a, we should move to all neighboring nodes through the links to them, setting in these nodes a distance from the start equal to the weight of the passed link, also naming the start node as their predecessor in an SPT.

Then we should move from these reached nodes in a similar way to their own neighbors, carrying into them the distance from the starting node, which is the sum of the distance brought to the predecessor node and the weight of the new link passed, also naming the previous node as their predecessor in SPT. This should take place only if the distance brought is smaller than the one already associated with the nodes, thus redefining it and the predecessor in these nodes (or if these nodes have been reached for the first time), otherwise we should stop—in these nodes only.

We should continue this throughout the whole network until possible. This spatial process easily (and always) converges to a solution, as in each step we either (a) come to the nodes for first time or (b) bring into them a shorter distance than before or (c) stop—and this cannot last indefinitely. Terminating globally, we will have one of the

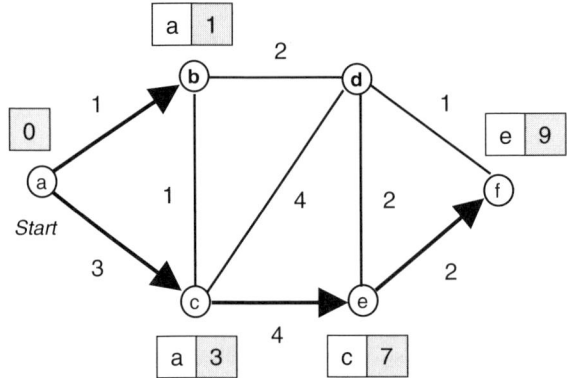

Figure 1.6. Initial SPT steps.

possible SPT with its links fixed in the network nodes by naming their final SPT predecessors.

SPT Finding Snapshots. Some initial development of the SPT finding process is shown in Figure 1.6. Starting from node a and propagating through its links, nodes b and c have been reached with setting distance in them from the start equal to the weights of links passed (shaded boxes) and establishing same predecessor a for both (clear boxes at the left of the shaded ones).

Further on, allowing arbitrary development of processes, let us assume that links from node c to node e, and then from e to f were passed, establishing corresponding distances from the start (as sums of weights of the passed links) and predecessors for the nodes reached.

An intermediate stage is shown in Figure 1.7, where distances from the start and predecessors for nodes c, e, and f have been redefined, as compared to Figure 1.6,

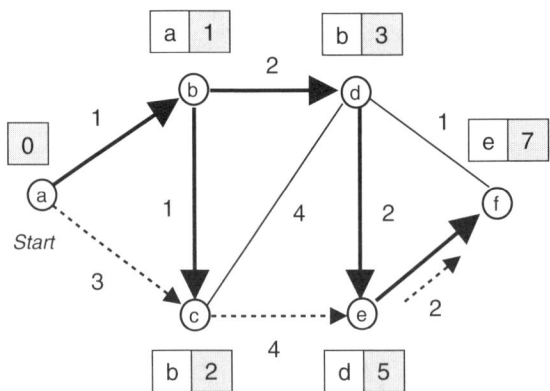

Figure 1.7. Intermediate SPT finding stage.

EXAMPLE: THE SHORTEST PATH PROBLEM

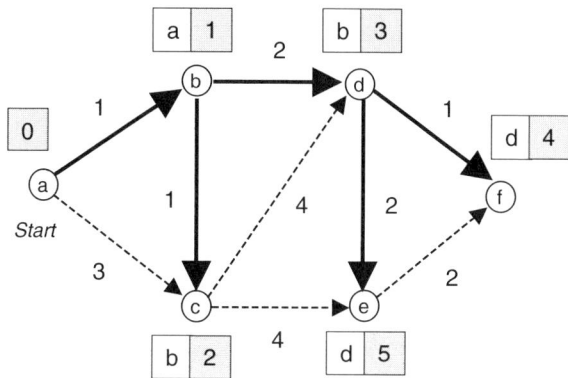

Figure 1.8. Final SPT solution.

with the previous path a-c-e-f discarded as not optimal. The link between nodes e and f was passed again, carrying shorter distance from the start and reinstating predecessor e in node f.

The final solution for the SPT (which in general may be one of the possible shortest path trees from the same start node) is shown in Figure 1.8. In comparison with Figure 1.7, an attempt to bring a shorter path to node d from node c failed, and the move from node d to node f redefined the distance and predecessor (as, correspondingly, 4 and d) in the latter for the final, optimal solution.

As follows from Figure 1.8, the resultant shortest path tree is recorded directly on the body of the network in a fully distributed form, by naming the SPT predecessor node in each of its nodes.

Spatial SPT Finding Pattern. A spatial pattern expressing the semantics, essence, of these shortest path tree finding processes, omitting any implementation details, is shown in Figure 1.9.

The pattern reveals maximum possible parallelism for solving the SPT problem, where all links from the nodes reached can be passed simultaneously, and operations in nodes do not need any synchronization with processes in other nodes that may cause delays. The pattern also characterizes an absolute distribution of SPT finding activities in the network space, and all operations it offers are purely local, being associated exclusively with the nodes reached.

A variety of concrete algorithms for sequential or parallel, as well as centralized or distributed, implementation can be designed and realized, all based on this most generic, semantic representation of the problem.

1.4.3 Collecting the Shortest Path Between Nodes

Tracing the Recorded Predecessors. To collect the shortest path from node a (from which we have already built SPT leading to all other nodes) and node e, we may

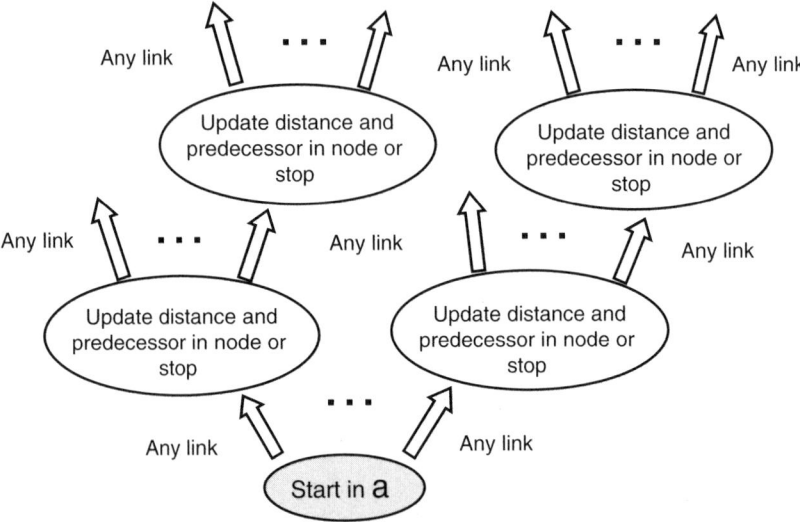

Figure 1.9. Spatial pattern of finding SPT.

start from node e and go to the neighbor node, which is its SPT predecessor, and then form the node reached to its own predecessor, and so on, unless we come into the node for which no SPT predecessor has been defined (and this will be node a) as shown in Figure 1.10.

During this movement between the network nodes by the recorded SPT predecessors in them, we can collect their names in the proper order and output the resultant path upon reaching the root of the tree, which is node a.

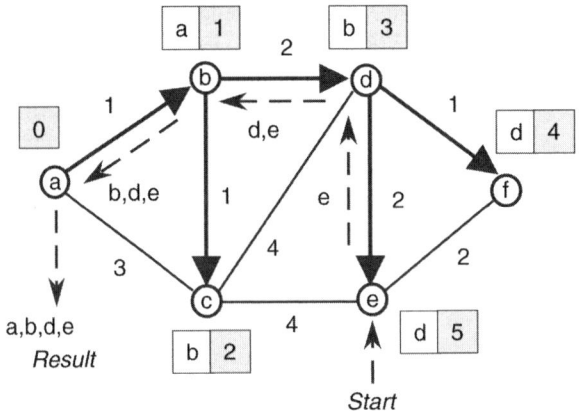

Figure 1.10. Collecting shortest path from node a to node e.

EXAMPLE: THE SHORTEST PATH PROBLEM

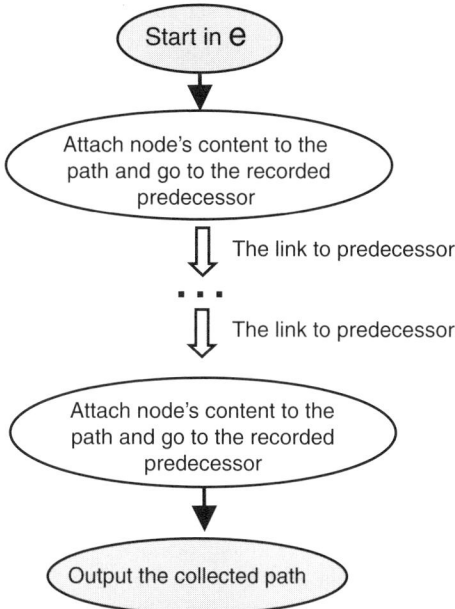

Figure 1.11. Spatial pattern for collecting the shortest path.

Shortest Path Collection Pattern. The spatial pattern expressing general semantics of this shortest path collection by moving between nodes using their prerecorded SPT predecessors is shown in Figure 1.11.

Similar to the SPT finding pattern of Figure 1.9, it may have different implementations in centralized or distributed environments too, all following, however, its sequential general nature.

Different algorithms can be possible to follow the semantics of finding shortest paths in networks, described above and expressed in the spatial patterns of Figures 1.9 and 1.11. For a sequential single-machine implementation, the first pattern has to be "sequentialized," and multidimensional arrays representing the network topology and different pointers in it can be used as data structures. In general, a single-machine implementation is trivial as all resources are in the same memory, and in view of the same program having an absolute control over the whole solution process.

1.4.4 Main Problems of Distributed Implementation

If we have a fully distributed representation of the network space of Figure 1.5, where each node may be in a different computer and links may happen to be between the computers, the shortest path solutions will be complicated drastically if traditional ideologies and technologies are used.

The following is only a fraction of what should be done in nodes of the network in addition to the operations exhibited in spatial patterns of Figures 1.9 and 1.11. Some of these service operations may be rather complex and may themselves need global supervision of the network.

- Loading the shortest path algorithm into *all nodes* of the network, checking the loading completion.
- Global search for the nodes by their names throughout the whole distributed network (here nodes a and e), starting sequentially two different distributed processes from these nodes, with assigning identities to these processes.
- Passing messages between nodes, interrupting processes for the data exchanges as these cannot be scheduled in advance in a distributed asynchronous system (the messages can come from any neighbors at any time and simultaneously from more than one neighbor).
- Broadcasting messages to all neighbors carrying distance from the start and the predecessor network address.
- Queuing messages coming simultaneously from different nodes, queuing replicated messages to be sent to successor nodes.
- Determining the overall termination of the SPT distributed process, in order to launch the path collection process, which can work only after the SPT is found.
- Informing the whole distributed network that the task is completed and the network should be cleaned up (i.e., variables for the path length and predecessor node address, still kept in all network nodes, should be removed) to be ready for execution of other distributed algorithms.

So we may be having something even more complex and ugly than what is exhibited in Figure 1.3, where parts 1 to 5 may be similar to nodes a to f of Figure 1.5, and additional control centers may need to be set up to solve the distributed coordination and management problems on different stages of the distributed shortest path solutions.

1.4.5 Universal WAVE-WP Interpreters

In WAVE-WP, all the routines listed above for distributed solutions are carried out automatically, with highest possible efficiency (as this is done on the internal system implementation, rather than application, level) by networks of the interpreters from WAVE-WP language. The interpreters may be associated with each node of the network, as shown in Figure 1.12, and reside in different computers, and the computers may have any communication infrastructure (which, e.g., can match the one of Figure 1.5 or can be quite different from it).

In WAVE-WP, the network of Figure 1.5 can also be arbitrarily distributed between the interpreters, with a possible partition between two interpreters (and two

EXAMPLE: THE SHORTEST PATH PROBLEM

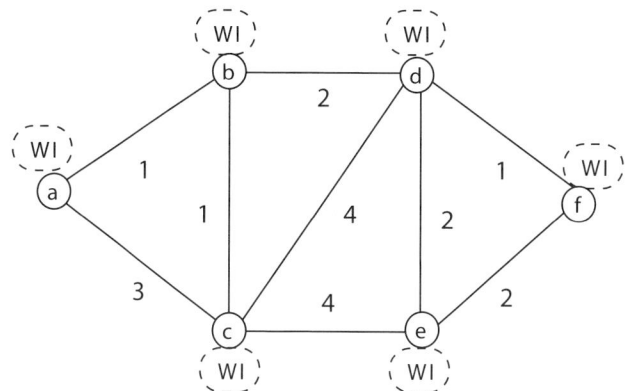

Figure 1.12. Associating the WAVE-WP interpreters with each node.

computers as well) depicted in Figure 1.13, with some links connecting nodes in the same interpreters while others being between nodes located in different interpreters (computers).

Direct Expression of the Semantics in WAVE-WP. In WAVE-WP, we can express the semantics of highly parallel and distributed solutions directly, omitting tedious system details symbolically exhibited in Figure 1.3 and also clarified above in relation to the shortest path problem.

For example, we can *straightforwardly* convert the spatial patterns of Figures 1.9 and 1.11 into active programs in WAVE-WP language dynamically self-navigating, self-replicating, and self-matching the distributed network in a pursuit of the global optimization solution, as follows.

1.4.6 Shortest Path Tree Finding in WAVE-WP

For finding a possible SPT in a network of weighed links in a maximum parallel and fully distributed mode, starting from node a and directly expressing the spatial pattern of Figure 1.9, the following WAVE-WP program will be sufficient:

```
direct # a;
repeat(
  any # noback; Fdistance += LINK;
  Ndistance == nil, Ndistance > Fdistance;
  Ndistance = Fdistance; Npredecessor = BACK
)
```

We will give below only some general explanation of how this program works, as the WAVE-WP language syntax and semantics are described in full details later in the book.

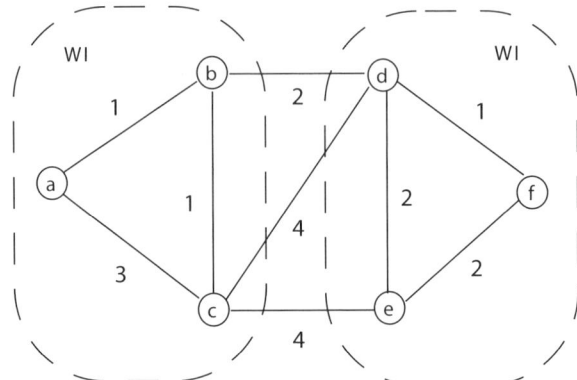

Figure 1.13. Partitioning the network between two interpreters.

The operations separated by a semicolon are executed in a sequence, while a comma separates parallel or independent parts. Symbol # is a hop act allowing us to move between nodes of the network, with the left operand giving information about links to pass and the right one detailing nodes to which these links should lead. This act permits both selective and broadcasting propagation in networks.

The operations of this program have the following meanings (given in the order written).

`direct # a`. Hops directly into node a of the network from an arbitrary point of the (distributed) world where this program is applied (which may also be any of the network nodes in Fig. 1.5).

`repeat(...)`. The `repeat` (belonging to nonlocal language constructs called *rules*) allows us to make spatial looping of the embraced part of the program, where the next iteration of this part starts in parallel from all nodes reached by the previous iteration—until such nodes exist.

`any # noback`. Makes hop from the current node (at the start this is node a) through *any* existing links *to all neighboring nodes, excluding the node it has been reached from*, using corresponding keywords as the left and right operands. The following part of the program will replicate, starting in parallel in all such nodes reached, if any.

`Fdistance += LINK`. The variable `Fdistance` (its name starting with F) belongs to the class of spatial variables called *frontal*. Being an exclusive property of the navigation program, this variable spreads and replicates during the space navigation (each program branch operating with its own copy of the variable).

`Fdistance` accumulates and carries the distance from the start node a, as the sum of weights of the links passed by the current branch. Having come into a node, this

EXAMPLE: THE SHORTEST PATH PROBLEM

variable (automatically created by the first access to it) is incremented by the content (here weight) of the link passed, which is accessible by special variable `LINK` (belonging to the class of *environmental* variables).

`Ndistance == nil, Ndistance > Fdistance`. These two statements, operating independently from each other, check if the variable `Ndistance` (belonging to N-prefixed *nodal* variables associated with network nodes) is undefined (treated as `nil`) or has the value that is greater than the one brought to the current node in variable `Fdistance`. If one of these statements results with `true`, the following part of the program will work. Otherwise the program in this particular node will stop (with the node failing to be classified as *reached*), not influencing, however, its development in other branches operating in other nodes.

`Ndistance = Fdistance; Npredecessor = BACK`. The sequence of these two statements redefines (or sets up if undefined, i.e., if the node is visited first time) the distance from the start in nodal variable `Ndistance` by the content of `Fdistance` brought into the current node. It also redefines (or sets up) in `Npredecessor` the network address of the predecessor node (lifted by environmental variable `BACK` in the current node). Nodal variables `Ndistance` and `Npredecessor` are automatically created by the first access to them (which is common to variables of WAVE-WP).

1.4.7 Shortest Path Collection in WAVE-WP

For collecting the shortest path using the SPT predecessors recorded in each node of the network in variable `Npredecessor`, while directly expressing the pattern of Figure 1.11, the following program will be sufficient:

```
direct # e;
repeat(Fpath = CONTENT & Fpath; any # Npredecessor);
USER = Fpath
```

Similar to the previous program, we will consider its work by analyzing different operations too, in the order written, as follows.

`direct # e`. Will hop directly to node `e` from some initial position in the world, which may be the same one we hopped from to node `a`, or any other.

`repeat(...)`. This spatial loop construct works similarly to the loop of the previous program. If the current iteration terminates with failure, the rest of the program (following the whole `repeat` construct) will work from the current node (i.e., the one from which this last, failed iteration started).

`Fpath = CONTENT & Fpath`. This statement adds the content (same as name) of the reached node (lifted by environmental variable `CONTENT`) to the beginning of the path being collected in frontal variable `Fpath` (the latter, automatically created on the first access to it, is moving between nodes too, accumulating the shortest path).

`any # Npredecessor`. Using the SPT predecessor node address recorded in nodal variable `Npredecessor` by the previous program, this statement hops to this predecessor node via any existing link to it. If no predecessor is recorded in a node (and this may only be the start node a), control state `fail` will result, terminating the spatial loop and also allowing the next operation to take place.

`USER = Fpath`. This will output the result by assigning the collected path in frontal variable `Fpath` to a symbolic user (using environmental variable `USER`), which may be any terminal inside or outside the WAVE-WP system or belonging to any other system.

1.4.8 Full Program for Finding Shortest Path

We can easily integrate the above two programs (for finding SPT from node a and collecting the shortest path from a to e using this SPT) within a single program starting from any point of the distributed world, as follows:

```
sequence(
  (direct # a;
  repeat(
    any # noback; Fdistance += LINK;
    Ndistance == nil, Ndistance > Fdistance;
    Ndistance = Fdistance; Npredecessor = BACK
  )
  ),
  (direct # e;
    repeat(Fpath = CONTENT & Fpath; any # Npredecessor);
    USER = Fpath
  )
)
```

Similar to the predecessor language WAVE, we can use abbreviations of the WAVE-WP language keywords (also short names of variables) to make WAVE-WP programs extremely compact and suitable for direct interpretation, without any compilation, in both software and hardware, as well as for quick spreading via communication channels between computers in a network. For the previous program, its short representation may be as follows:

```
sq(
(@#a;rp(#nb;F+=L;N==,N>F;N=F;Np=B)),
(@#e;rp(F=C&F;#Np);U=F)
)
```

DISTRIBUTED KNOWLEDGE REPRESENTATION

Using some additional features of WAVE-WP, this program can be made even shorter, for example, as:

```
@#a;[rp(#;F+=L;N==,N>F;N=F;M=B)];
@#e;rp(F=C&F;#M);U=F
```

It essentially is (up to two, or even three, orders of magnitude!) shorter than in any other languages, being at the same time fully distributed and maximum parallel solution of the shortest path problem.

In WAVE-WP, we can express the semantics of spatial problems directly, ignoring implementation details, delegating the latter to an efficient automatic interpretation system. This is the main distinction and advantage of WAVE-WP paradigm in comparison with any other models and technologies.

1.5 EXAMPLE: DISTRIBUTED KNOWLEDGE REPRESENTATION AND PROCESSING

In the previous section, we made an attempt to explain what WAVE-WP means on an example of the important computation and optimization problem on distributed networks. Let us now try to touch another very important direction: *knowledge representation and processing*, especially parallel and distributed ones, which underlies a great variety of intelligent systems of different natures and with diverse applications.

1.5.1 Knowledge Network

The most general knowledge representation may be in the form of a knowledge network, or KN, with nodes reflecting different concepts or facts, and links being relations (oriented as well as nonoriented) between the nodes, with an example shown in Figure 1.14. In this KN, persons named `Masanori`, `Peter`, `Bob`, `John` and `Doug` are shown in their various relations (symmetric as well asymmetric, directed) to each other and also to such substances as `whiskey` and `beer`.

There are different strategies for creating any knowledge networks in parallel and fully distributed mode in WAVE-WP, inheriting these from its predecessor WAVE. Very efficient and compact ones usually based on depth-first tree templates covering the network.

Without explaining the details in this introductory presentation (which may be easily found in the previous book as well as later in this one), we are showing below how the WAVE-WP program creating the KN of Figure 1.14 in a "single breath" may look like. (The symbol ## identifies hops to the already existing nodes, up the depth-first tree, and the language rule `create` supplies the program it embraces with a global creative power when navigating in space).

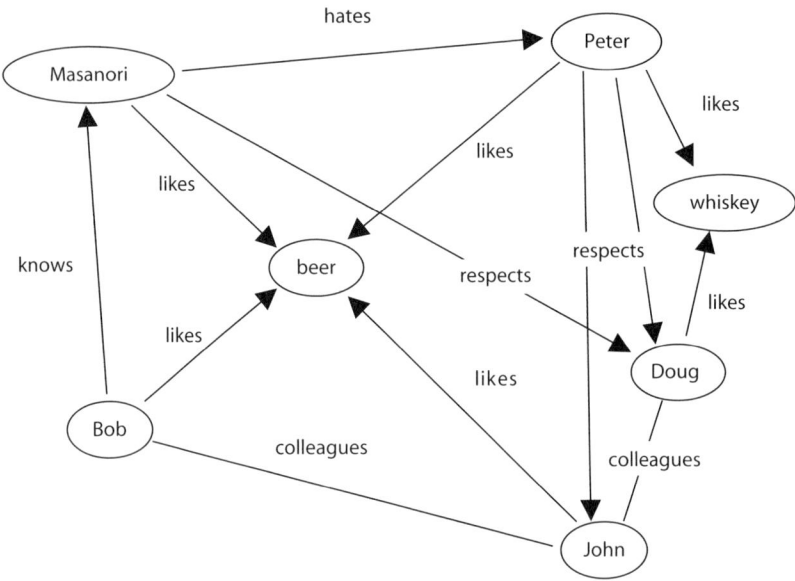

Figure 1.14. A knowledge network.

```
create(
  direct # masanori; + likes # beer; - likes # john;
  (colleagues # bob; + likes ## beer, + knows ## masanori),
  (colleagues # doug; - respects ## Masanori,
    (+ likes # whiskey; - likes # Peter;
    + respects ## (john, doug), + likes ## beer,
    - hates ## masanori)))
```

During the creation, this network may be arbitrarily (or in any other needed way) distributed between interpreters installed, say, in robots, with a possible partition shown in Figure 1.15. We can also easily supply the program with explicit hints of which hardware or software resources should be used for the implementation, or even with a detailed mapping on particular hardware (robots, in our case).

Let us show a couple of examples of elementary tasks that can be solved on this created KN in a spatial pattern-matching mode, which is effectively provided by WAVE-WP (as well as by its predecessor WAVE).

1.5.2 Elementary Query 1

"Find everybody who likes beer"

The spatial pattern for this request is shown in Figure 1.16, where the solution to be found corresponds to nodes named X.

DISTRIBUTED KNOWLEDGE REPRESENTATION

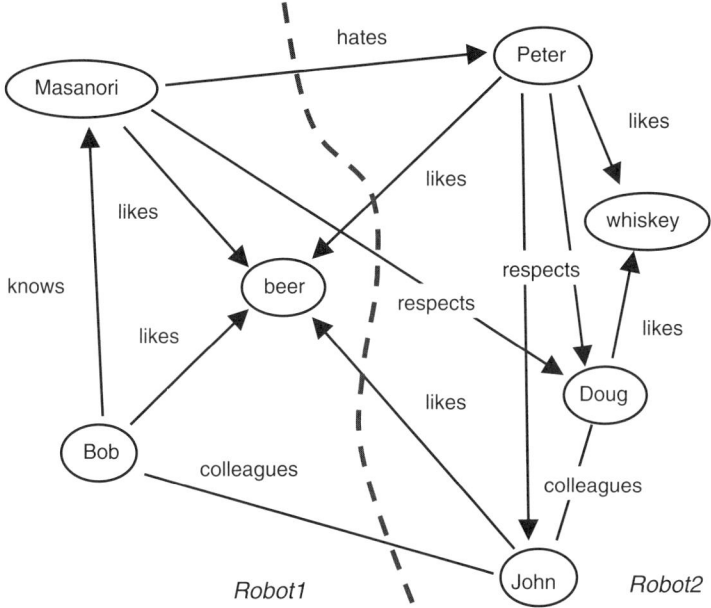

Figure 1.15. Possible network distribution between two interpreters.

In WAVE-WP, we can write a program straightforwardly following this pattern and providing on output of the contents (names) of all X-type nodes reached in a hop from node beer opposite all oriented links named likes.

```
USER = (direct # beer; - likes # any)
```

So in WAVE-WP the pattern of Figure 1.16 converts into a self-navigating parallel program that does all the job by itself, including the output of the (possibly, multiple) result. Starting from node beer, the program above finds the four resultant nodes: bob, john, masanori, and peter, as shown in Figure 1.17 (the nodes matching the pattern are shaded and matching links are shown in bold).

These nodes are found in a parallel broadcast from node beer and happened to reside in different robots, with bob and masanori in Robot1, and john and peter

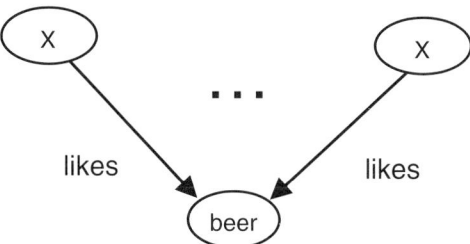

Figure 1.16. Spatial pattern for query 1.

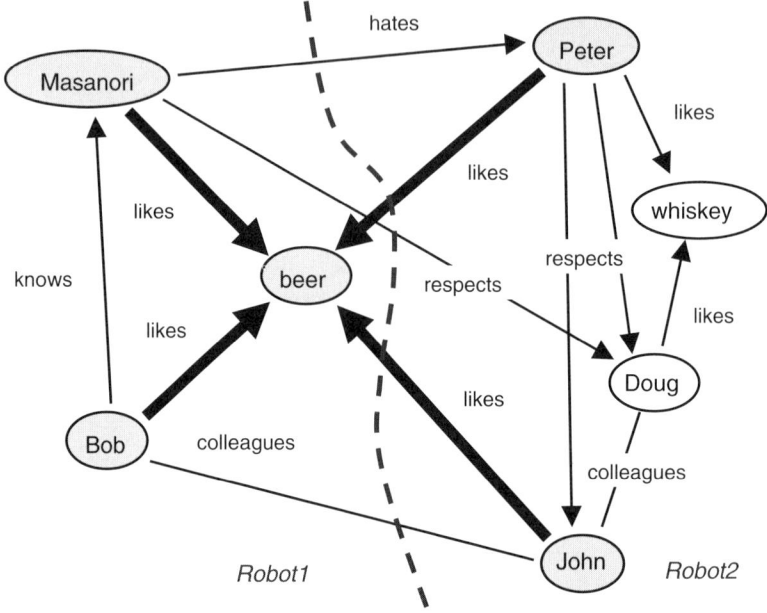

Figure 1.17. Spatial solution for query 1.

in Robot2, but their names are collected in one list and output in the world point where the pattern was applied, which can be any location, including one of the robots engaged.

1.5.3 Elementary Query 2

"Who from John's colleagues is respected by both Masanori and Peter?"

This request corresponds to the spatial pattern of Figure 1.18, which is to be matched with the KN of Figure 1.14 (or 1.15) to provide the solution needed.

This pattern can, too, be easily converted into an active WAVE-WP program solving the task in parallel and fully distributed mode, regardless of the KN distribution between interpreters (robots), as follows:

```
direct # John; colleagues # any;
and(
  andparallel(- respects # (Masanori, Peter)),
  USER = CONTENT
)
```

Starting from node john, the program moves from it in parallel via all links named colleagues, and checks independently in the nodes reached (i.e., bob and doug) whether they have oriented links respects coming *from both* peter and masanori

DISTRIBUTED KNOWLEDGE REPRESENTATION

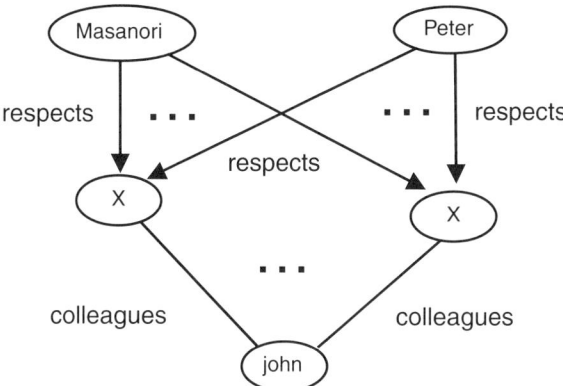

Figure 1.18. Spatial pattern for query 2.

nodes. The program provides the final single answer doug found in Robot2, as shown in Figure 1.19, which is copied and sent outside (using CONTENT and USER environmental variables).

The program uses and as well as andparallel (parallel and) rules of the language providing the needed spatial logic in finding the distributed solution. The statement:

andparallel(- respects # (Masanori, Peter))

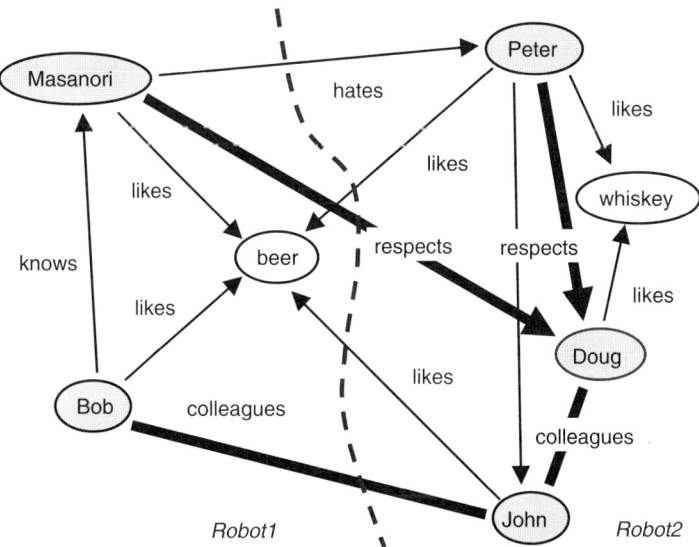

Figure 1.19. Spatial solution for query 2.

will be automatically (due to the WAVE-WP language syntax and semantics) split into:

```
andparallel(- respects # Masanori, - respects # Peter)
```

The whole program works regardless of the KN distribution between the interpreters (robots), which may vary from associating of a separate copy of the interpreter with each KN node (like what we have shown for the network in Fig. 1.12), to keeping the whole KN within the single interpreter, thus degenerating into a single-machine solution, as usual.

1.6 SYSTEM ORGANIZATION AS A FUNCTION OF THE APPLICATION SCENARIO

As is demonstrated by the above programming examples, WAVE-WP allows us to express the semantics of distributed problems straightforwardly, in a very compact and clear way, abstracting from the system implementation details. At the same time, the space-navigating algorithms in WAVE-WP are highly parallel and maximum distributed in nature, requiring conceptually no central resources, thus providing the impetus for most efficient implementations in distributed environments.

The overall system organization, similar to what has been shown in Figure 1.3, may not be predetermined in advance in our case, but may rather represent a *function of the application scenario*, the latter written on the layer well above the traditional partitioning into system components and their communications.

Materialization of the system executing these high-level spatial scenarios may be provided at run time and may depend on the current external environment in which these scenarios operate, as well as on the currently available software and hardware resources. The automatic implementation may support the integrity of these scenarios and their ability to behave as a whole under any circumstances, despite failures or damages of the system components.

Having radically raised the distributed problem description level, we have effectively shifted the extremely complex and tedious routines of the internal system organization and overall management and control to the efficient automatic layer, as will be clear from of rest of this book.

1.7 RELATION TO THE PREVIOUS BOOK

In the previous book (Sapaty, 1999a) a new distributed processing model, WAVE, was introduced. It allowed us to dynamically create virtual, or knowledge, networks in distributed computer networks and process them in parallel by multiple navigating processes.

Being just a formal spatial graph machine, with graphs universally representing any knowledge and systems, WAVE was capable of solving complex problems in computer networks from many different classes. The solutions were integral, seamless, and compact due to higher level programming in WAVE, with traditional distributed control and management routines performed automatically on the language interpretation level.

In WAVE-WP, along with more advanced knowledge processing in open computer networks, which is also done with the help of knowledge networks, we can move directly in a continuous physical world and organize multiple activities there, including physical matter (or objects) processing. The implementation can be performed fully automatically too, by multiple mobile manned or unmanned hardware.

By using a similar to WAVE spatial automaton matching now continuous physical worlds too, this enables us to cover much broader and very important areas (practically all) of working in real environments, for both humans and robots. This includes, for instance, organization and management of human collectives, cooperative behavior of groups of robots (up to robotic armies), as well as cooperation and symbiosis between humans and robots in the advanced society—all as a derivative of high-level scenarios expressed in WAVE-WP.

1.8 COMPARISON WITH OTHER WORKS IN RELATED AREAS

For the 5 years after the publication of the first book on WAVE (Sapaty, 1999a), this new paradigm, having evolved now into a much broader and more powerful WAVE-WP, still remains unique, neither matching other approaches nor falling into other categories or classes. It, however, has features and applications that may intersect with other areas. Some possible links are outlined below.

1.8.1 Parallel Computing

WAVE-WP strongly relates to parallel computing [see, e.g., Gupta et al. (2003) and Dongarra et al. (2002)] as it allows spatial processes to develop with maximum freedom and minimum ordering, revealing all potential parallelism for any implementation. The spatial algorithms in WAVE-WP often resemble the spread of smog or flooding, and they are often more natural, simple, and compact than the related ones programmed in conventional languages, Java, for example. Of course, there may be cases where spatial constraints like synchronization and ordering are vital, and WAVE-WP has very efficient spatial mechanisms for providing these, but only if absolutely necessary.

1.8.2 Distributed Systems and Distributed Computing

WAVE-WP fully orients on distributed systems and distributed computing (Attiya, 2004; Bruce and Dempsey, 1996; Tanenbaum, 2002). Representing a completely different paradigm from conventional terms, it allows us to design fully distributed algorithms where any elementary operation and any locus of control are always associated with local points in distributed physical or virtual spaces. This brings the distributed computing (processing) philosophy to its extreme, keeping at the same time the highest integrity of distributed solutions comparable to single-machine implementations.

1.8.3 Parallel and Distributed Computing

Integration of the ideas of parallel and distributed computing, especially with the efficient use of distributed computer networks, and for such important applications as, say, distributed interactive simulation (Hughes, et al., 2003; Fujimoto, 1999), is of growing interest and importance. In WAVE-WP these two directions integrate most naturally and are inseparable from each other, as parallelism naturally appears during the code replication and free spreading in a distributed world.

1.8.4 Computer Networking

Computer networking (Comer, 2000; Tanenbaum, 2002) is another strong link with WAVE-WP, as originally this paradigm, through WAVE, was designed to work in distributed and open network spaces and was effectively used for distributed network management; see Sapaty (1999a). Any network algorithms and protocols can be expressed in WAVE-WP as in the universal spatial automaton. Fighting crisis situations in computer networks, especially withstanding virus attacks, is one of the promising applications of WAVE-WP, to be discussed later in the book.

1.8.5 Intelligent Agents

Design of intelligent agents and efficient composition of complex systems from them (Bigus et al., 2001; Wooldridge, 2002) is one of the hottest topics in computer science and artificial intelligence. In WAVE-WP, a copy of the language interpreter, which can communicate with other such copies, is actually a universal programmable agent, and a static or dynamic network of such interpreters forms a spatial machine capable of executing any distributed algorithms represented in WAVE-WP.

1.8.6 Mobile Agents

Mobile agents (Earnshaw and Vince, 2002; Lange and Oshima, 1998; Milojicic et al., 1999; Siegwart et al., 2004) allow us to move through the distributed network and perform operations and collect data remotely. WAVE-WP, too, allows us to navigate distributed systems and execute operations where data resides, but this is organized on a much higher level than traditional mobile agents, without dividing the program into agents and their interactions.

Such things, equivalent to traditional mobile agents and also advanced distributed command and control over them (even unthinkable to be done using traditional mobile agents) are done exclusively on the implementation layer, allowing us to have application programs hundreds of times simpler and shorter that if written on the level of mobile agents directly.

1.8.7 Grid Computing

The grid computing (Carey, 1997; Abbas, 2004; Foster and Kesselman, 2003; Joseph and Fellenstein, 2003), inheriting the idea from electrical grids, tries to organize a shared use of all distributed computational resources, not only distributed information. WAVE-WP, providing this too but much cheaper and also more efficient, allows us to set up a higher level of the system organization. It converts the whole world into a powerful spatial machine that may be effectively driven by a single spatial program, dynamically covering the whole computerized world, if needed.

Embedding WAVE-WP interpreters into sensitive points of the artificial or natural distributed systems, we may obtain a unique possibility to coordinate, or *rule*, these systems in any needed way, with grid computing being years away from these, already existing, global possibilities.

1.8.8 Spatial Programming

The term *spatial programming* has been recently used for programming computers embedded into the physical world (Borcea et al., 2004; Iftode et al., 2003). Separately from this, *spatial* also relates to handling information with a geographical element (Wood, 2002). WAVE-WP, supporting all these features too, also allows us to program arbitrary complex operations in physical, virtual, or combined worlds *directly*, abstracting from computers and their networks. We also say that WAVE-WP programs are "spatial" in nature, meaning they *exist and evolve in space* (like, say, holograms) rather than in particular points. Throughout this book, we will be using the term in this broader sense, as a key feature of the programming philosophy.

1.8.9 Mobile Robotics, Cooperative Robotics

Allowing us to move in physical worlds directly and process both information and physical matter there, WAVE-WP, unlike WAVE has become very close to the topic of mobile robots (Braunl, 2003; Dudek and Jenkin, 2000; Holland, 2003) on the implementation level, where the language interpreters can be placed on top of the robotic functionality. And their ability to form a universal distributed computer driven by high-level scenarios in the spatial language allows us to cover with this philosophy any distributed multirobot systems (Asama et al., 2002; Butenko et al., 2003; Parker et al., 2000) and their collective or cooperative behavior.

1.8.10 System Management

Complex system management (Laudon et al., 2003; Wheelen et al., 2003) is one of the main applications of distributed WAVE-WP control model and technology, and the rest of this book is mainly devoted to the organization and management of a variety of large dynamic and open systems.

1.9 ORGANIZATION OF THE BOOK

In Chapter 2, key features of the WAVE-WP model are considered with examples and illustrations. Three distributed worlds can be integrated within the same formalism: continuous physical world, discrete virtual world, and the world of communicating doers, or execution world. The united world can be navigated by parallel spatial programs, *waves*, which can establish any authority over it.

In Chapter 3, the WAVE-WP language is described. Detailed syntax and semantics of the language are presented along with elementary programming examples explaining them. WAVE-WP operates on values that may represent information, physical matter, or any combination of the two. It is a higher level language for spatial processing and control, which can directly access any points of distributed worlds, process local or remote data, as well as assign local or remote results to local or remote variables.

In Chapter 4, we consider the peculiarities of distributed interpretation of WAVE-WP language by dynamic networks of "doers" represented by the language interpreters embedded in communicating vehicles of any origin. The solutions discussed can allow spatial WAVE-WP scenarios to evolve in various environments and set up advanced command and control over other systems, with automatic decision making at any levels.

Chapter 5 contains examples of the practical use of WAVE-WP language ranging from traditional programming to expressing higher level abstractions and cooperative actions in the united physical–virtual space, with implementation in computer networks and multiple mobile robots. Included are elements of the new agent-free methodology of distributed and parallel programming of dynamic systems in a spatial pattern-matching mode.

In Chapter 6, an expression in WAVE-WP of a number of exemplary mission scenarios in a distributed physical world is considered. Different forms of movement of organized groups, including the movement with physical payload, are presented. Distributed space search and processing are discussed too. Optimized solutions can be found in the virtual world first in a look-ahead simulation mode, and subsequently used for guiding parallel movement and operations in the physical world.

In Chapter 7, we consider the run time creation and modification of different kinds of distributed command-and-control infrastructures for mobile dynamic systems, with orientation on crisis management, and also solving some basic control and management problems in a spatial mode, navigating these infrastructures in parallel. Rapid composition of an effective crisis reaction force and providing its high operability in hostile environments are extremely important problems to be solved.

Chapter 8, as a continuation of Chapter 7, concentrates on a variety of dynamic situations and scenarios that may take place in multirobot, computer network, and united global systems, especially with the relation to advanced crisis management. All these are investigated and expressed using the WAVE-WP language. Distributed cognitive systems, robotized hospitals, future combat systems, and fighting virus attacks in computer networks are among the topics and problems discussed, with examples of distributed solutions presented.

ORGANIZATION OF THE BOOK

Chapter 9 concludes the book, summarizing main features of the WAVE-WP model and technology, also highlighting a number of its most important applications at present as well as revealing future plans.

The Appendix contains full WAVE-WP language syntax in both extended and compact forms, as well as main abbreviations of the language keywords allowing us to make programs very compact.

The References list main publications on the WAVE-WP model (Sapaty, 1999b; 2000–2004) and also relate to further practical applications of the previous model WAVE (Gonzalez-Valenzuela and Leung, 2002; Gonzalez-Valenzuela and Vuong, 2002; Gonzalez-Valenzuela et al., 2001) and its implementation (Borst, 2002), which appeared after the first book was published. Other works in the related areas are included too.

2

WORLDS AND WAVES IN THE WAVE-WP MODEL

Key features of the WAVE-WP (or World Processing) model will be considered in an informal way, with examples and illustrations. The main difference between WAVE-WP and its predecessor WAVE (Sapaty, 1999a) is in the extended concept of the world with which it operates. Three distributed worlds will be considered in WAVE-WP within the same formalism: continuous physical world, discrete virtual world, and the world of communicating doers, or execution world (with doers being any computerized devices, humans included).

These distributed worlds can be spatially interlinked and deeply integrated with each other, and mappings between them may be dynamic, changing over time. The united worlds can be effectively accessed in a variety of ways, using combined possibilities from the constituent worlds.

The united worlds are navigated by parallel spatial programs, or waves, that can establish any authority over the worlds, including controlling, processing, modifying, and creating them, also influencing their further behavior and evolution. Waves in WAVE-WP have a higher expressive power than in WAVE, allowing higher level structuring of expressions and direct work with remote data, which may represent physical matter too, not information only.

Ruling Distributed Dynamic Worlds, by Peter S. Sapaty
ISBN 0-471-65575-9   Copyright © 2005 John Wiley & Sons, Inc.

2.1 PHYSICAL WORLD

2.1.1 Temporary Physical World Nodes

Physical world (or PW) is continuous and infinite in WAVE-WP. Existing in any of its point, and possibly performing a job there, is symbolically considered as staying in a *node*. Such a node, reflecting only occupancy (or occupancies) in the point, does not have any personal identity or content beyond this fact. The node disappears as soon as all occupancies in it terminate.

Any point (node) in PW can be represented by its (absolute) *coordinates* using which the point can be reached physically. Moving to other nodes can use coordinates of the destination points too or just changes (*shifts*) in coordinates from the current node. In a combined case, on some coordinates their absolute values and on others their shifts can be used. We will be using hereinafter absolute coordinate values prefixed by x, y, and z and their shifts prefixed by dx, dy, and dz, respectively.

Examples of some nodes in a two-dimensional space and moves between them by absolute coordinates or coordinate shifts (or combined) from the starting node are shown in Figure 2.1.

Nodes with coordinated x1 y7, x6 y6, and x11 y10 will terminate as soon as the needed hops from them to other PW locations are performed, forming new nodes at the destinations with coordinates, correspondingly, x9 y8, x12 y2, and x12 y6. The former nodes will, however, remain if they have other occupancies or activities associated with them (same can be said about nodes x4 y3 and x15 y5).

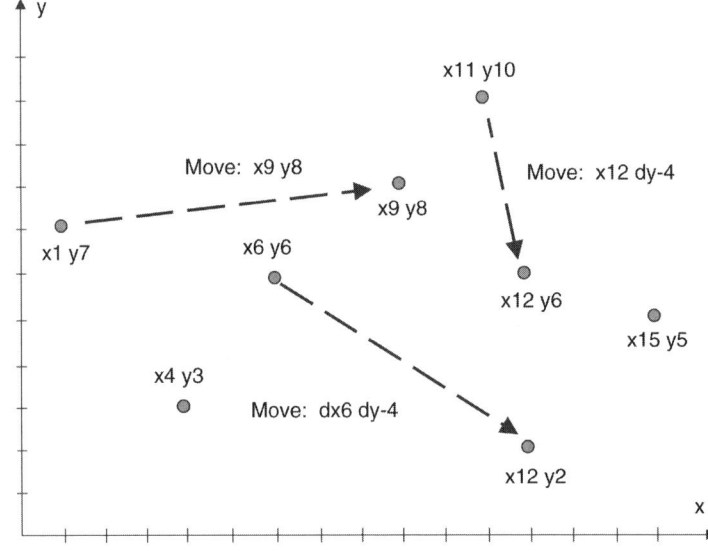

Figure 2.1. Temporary nodes in physical world and moves between them.

2.1.2 Visiting Existing Nodes in a Region

As PW is continuous, and coordinates always have limited precision, it is considered virtually impossible to reach exactly the same location more than once. Also taking into account that PW nodes reflect only the presence of activities in them, which are always unique in time, any new movement in PW is always assumed as forming new, unique, nodes.

It is nevertheless possible in WAVE-WP to visit the already existing PW nodes, formed and occupied by other activities. Reentering such nodes can bring to them additional activities accompanying the new visits, which will be allowed to use and share local data with other activities already in these nodes.

As it is not generally possible to come exactly to the same point in a continuous PW more than once by coordinates, as stated before, a certain *region*, rather than point, should be defined in PW where these nodes are expected to exist. This search region, which may be arbitrarily large (up to the whole world), can be specified by coordinates of its *center* and a *range*, or radius (with its value prefixed by r) from this center, as an additional spatial dimension, as is shown in Figure 2.2.

More than one node can be reentered in such a way, bringing to all these nodes the same new occupancy (and activity, if any). The region in Figure 2.2 is set up by its center x11 y7.5, and radius r4, and the three nodes with coordinates x9 y8, x11 y10, and x12 y6 lying in this region, will be reentered simultaneously in this operation. Reentering an existing PW node may prolong its lifetime, as the new occupancy will sum up with the occupancies already present there.

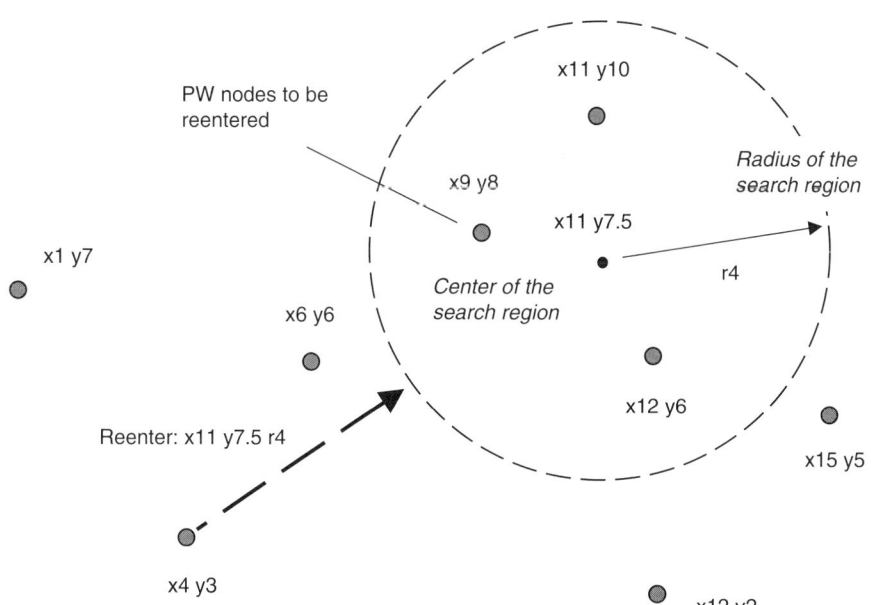

Figure 2.2. Reentering existing PW nodes lying in the region given.

2.1.3 Destination Regions for New Nodes

The idea of outlining a region rather than point to be accessed in the physical world may also be useful for the case of moving into new PW positions, where the given range may serve as an allowed precision (or deviation) of reaching the destinations defined by coordinates. This may, for example, relate to reaching remote places (say, on other planets) or to work in complex physical spaces (like rugged terrain with certain places unsuitable for landing, or rocky inclusions to be avoided during the earth tunneling). Arriving at a point outside the region allowed may be classified as the operation's failure, as shown in Figure 2.3.

2.1.4 Accessing Physical World Parameters

When staying in PW nodes, it may be possible to lift different parameters of the world in these locations (say, elevation, temperature, density, wind speed, chemical composition, humidity, radiation level, visibility, etc.) and by this investigate the world. It may also be possible to impact the physical world by attempting to change certain parameters in (or via) these locations. This dotlike influence, however, may seem problematic in general and may need introduction of special operations covering large enough regions as a whole, rather than using their selected points.

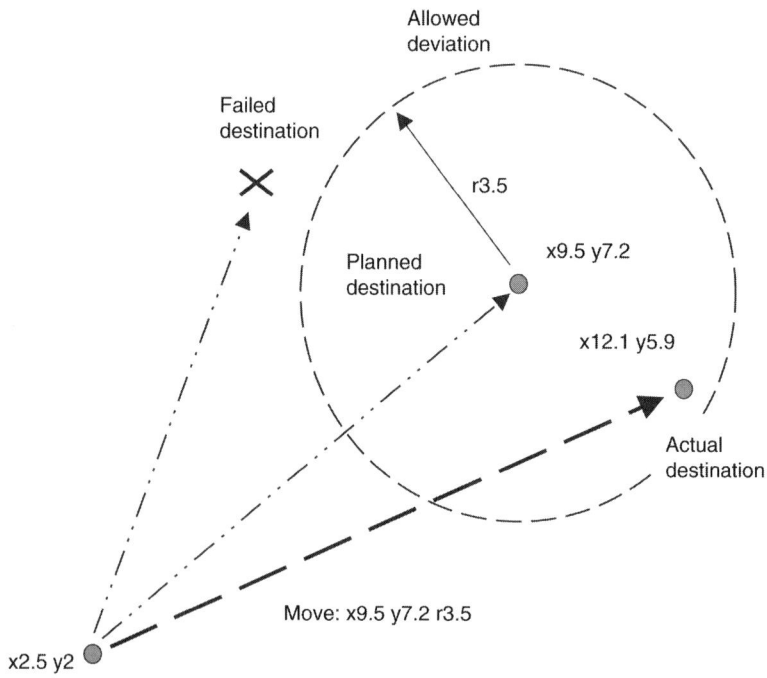

Figure 2.3. Defining an allowed destination region in physical world.

2.1.5 Broadcasting in Physical World

Staying in any node of the physical world, it is possible to move simultaneously to many other nodes, either to new ones with their creation or to the already existing nodes (as shown in Fig. 2.2), with *broadcasting* navigation in PW being one of the key features of the WAVE-WP model. To move into more than one new node, their absolute coordinates or coordinate shifts from the current node may be listed in a single operation, like what follows, in order to receive simultaneously all the nodes shown in Figure 2.1:

```
Move: x1 y7, x4 y3, x6 y6, x9 y8, x11 y10, x12 y6, x15 y5, x12 y2
```

This broadcasting movement in PW with parallel creation of the new temporary nodes is shown in Figure 2.4, being applied in some PW point (for which the one with coordinates x9 y4 is chosen).

When reentering the existing nodes, it is also possible to move by a single operation simultaneously to more than one region in PW (actually to as many as needed) and, correspondingly, to all nodes lying in these regions. If the regions intersect in PW, the nodes lying in the intersections can be reentered more than once, as shown in Figure 2.5 for the following operation (supposedly starting in node x4.5 y2.8):

```
Reenter: x4.5 y7.5 r3.5, x7.6 y8.5 r2.8, x11.5 y8.2 r3.2
```

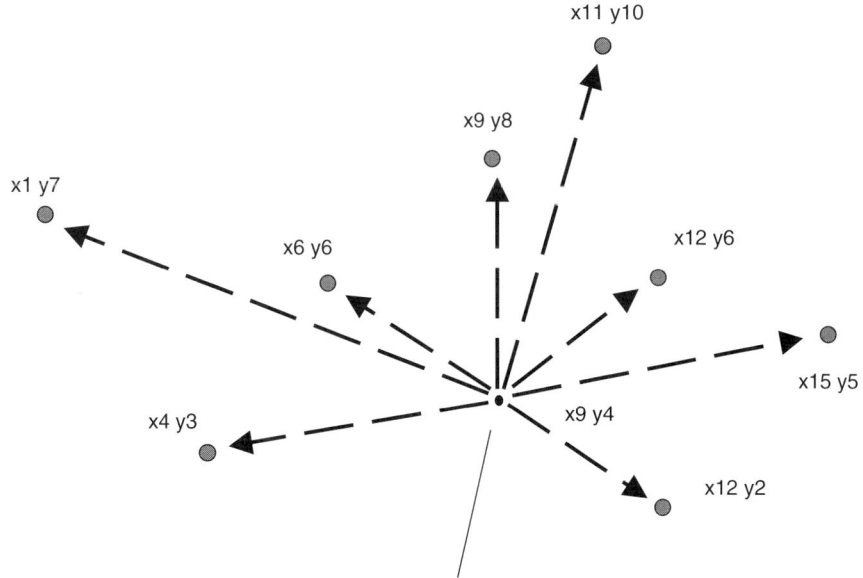

Move: x1 y7, x4 y3, x6 y6, x9 y8, x11 y10, x12 y6, x15 y5, x12 y2

Figure 2.4. Node-creating broadcasting in physical world.

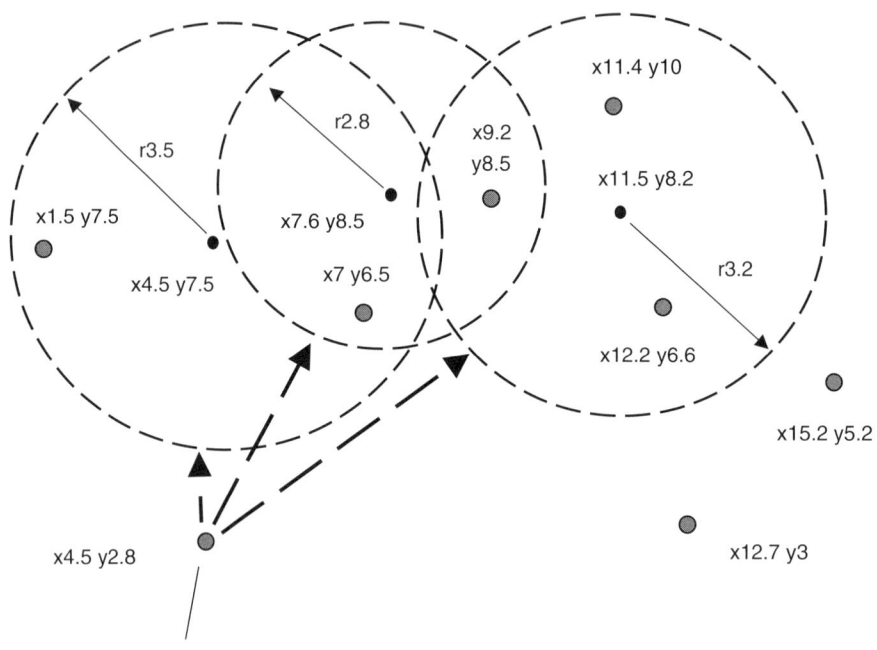

Figure 2.5. Multiple reentries of multiple nodes in physical world via intersecting regions.

The following nodes will be reentered: x1.5 y7.5, x12.2 y6.6, x11.4 y10, x7 y6.5 (twice), and x9.2 y8.5 (twice).

2.2 VIRTUAL WORLD

2.2.1 Knowledge Networks

Virtual world (or VW) is discrete and interlinked in WAVE-WP, and similarly to the WAVE model, and is represented by a knowledge network (KN). Its nodes may contain established concepts or facts, and links (oriented as well as nonoriented), connecting the nodes, may reflect different relations between the latter.

Virtual world nodes in KN, unlike PW nodes, represent themselves as they are and may exist regardless of presence or absence of occupancies or activities in them. They always have *contents*, same as *names* (what is more appropriate depends on the context of use). These may represent any information, as arbitrary long strings of characters (including programs in WAVE-WP or other languages to be executed). Links in KN may have arbitrary contents (names) too, same as nodes.

An example of a KN is shown in Figure 2.6, where node1 to node8 and link1 to link9 are the chosen names (contents) of nodes and links, which may be any other strings of characters (e.g., John, 'alpha + beta', or 667.09).

VIRTUAL WORLD

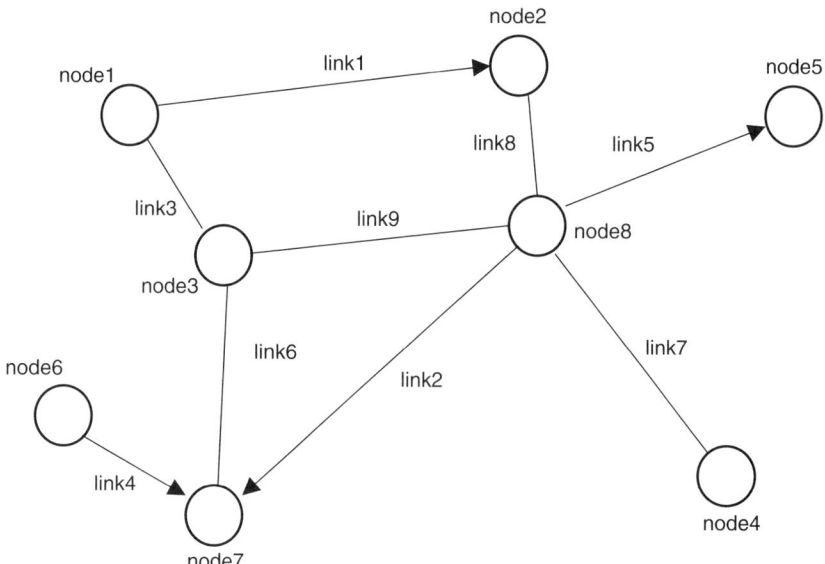

Figure 2.6. Representing virtual world by knowledge network.

Knowledge network topology can be arbitrary, and the network may be of any size. Number of nodes and links, their contents, and network structure may change over time, so KN can be dynamic and open.

The KN nodes and links are persistent: After creation, they exist for an unlimited period of time, unless removed explicitly. Creation of new nodes and links can be performed in parallel, when staying in the same or in different nodes. Creation of new parts of KN can be performed simultaneously with modification and deletion of the existing parts of it.

Removal of VW nodes is automatically accompanied by the deletion of all links adjacent to these nodes. The removal of nodes can be *graceful*, waiting until all activities in them terminate first (similar to the natural termination of temporary PW nodes after all occupancies in them terminate). KN node removal can also be *rough*, where everything associated with the nodes (including any activities and temporary results) is aborted immediately.

2.2.2 Access to Nodes and Links

Nodes of KN can be created, referred to, and reentered globally by using their contents (names), if known. More than one KN node may have same content, so referring to nodes by contents may simultaneously lead to more than one node in general.

Links of KN cannot be accessed globally by their names such as nodes, as all operations within the WAVE-WP model are performed in nodes. The main possibilities of dealing with certain links, which can be accessed from the adjacent nodes only, are as follows.

To move through particular links, they should be named ("contented") explicitly in the starting node; and simultaneous passing of a group of differently named links from the same node can be done by listing all their contents together within one hop operation.

The groups of links from a node can also be accessed without naming links individually by saying, for example, that *all incoming* or *all outgoing* (if they are oriented), or *all associated* links (with any orientation) should be passed from a node. As links adjacent to a node may have repeating contents (names), even naming links explicitly in a node may generally result in passing more than one link.

Lifting and (or) changing the link information (its content and orientation), as well the link deletion, is possible only *after* moving through the link—in the destination node. Such a link is classified and accessed in this node as the (just) *passed* one.

2.2.3 Tunnel and Surface Broadcasting

Movement in KN between nodes may be classified as direct (or *tunnel*), by node contents, and *surface*, via links to other nodes. Tunnel movement by node contents may generally lead to more than one node simultaneously (as nodes may have the same contents, as mentioned before), in the form of *local broadcasting*. Listing more than one node name in a tunnel hop may extend the local broadcasting possibilities, as shown in Figure 2.7, with hopping from `node1` simultaneously to `node2`, `node7`, and `node8`.

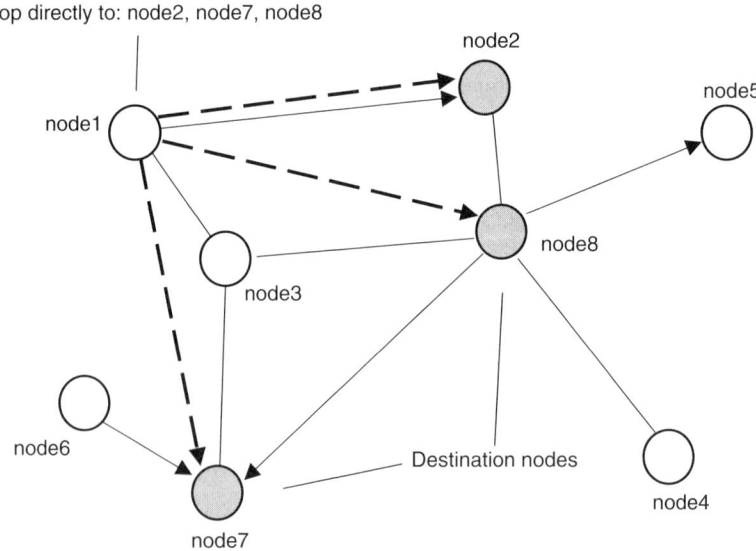

Figure 2.7. Local tunnel broadcasting from a node.

VIRTUAL WORLD

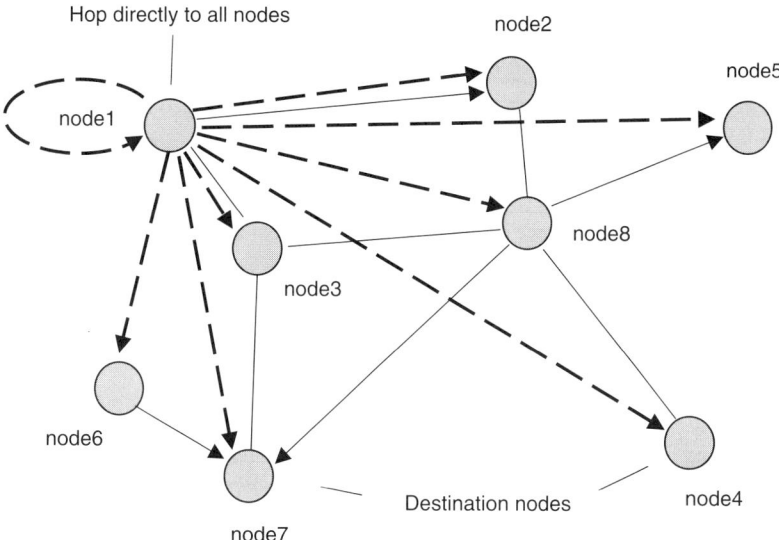

Figure 2.8. Global tunnel broadcasting from a node to all nodes.

Any number of node names (contents) can be given for this local broadcasting hop (up to all nodes of the network, including the starting one, making the broadcasting global).

The *global broadcasting* from a node to all KN nodes can also be organized without individually naming the destination nodes, by a special tunnel *hop to all* nodes of a KN, shown in Figure 2.8 (starting in node1 again).

This global broadcasting may have different options. For example, a hop *to all other* nodes (i.e., excluding the current, starting one) or hops to all or all other nodes *except the predecessor* node (the one from which the current node was accessed) may take place.

For the surface movement, naming links (individually or as a group) in the starting nodes may be combined with naming the nodes these links should lead to, making possible destinations more selective. Accessing groups of links in nodes may in general cause *local broadcasting* to a number of adjacent nodes. This is shown Figure 2.9 by moving from node8 simultaneously via link5, link7, link8, and link9 to, correspondingly, node5, node4, node2, and node3.

The VW nodes, along with links connecting them, are not bound to any positions in PW and can exist anywhere, being also free to migrate in the physical space, while always preserving the logical structure of the KN.

2.2.4 Linking with Alien Networks

The VW nodes and links may not only be the property of the WAVE-WP system itself but can also belong to other, "alien" systems existing independently, which

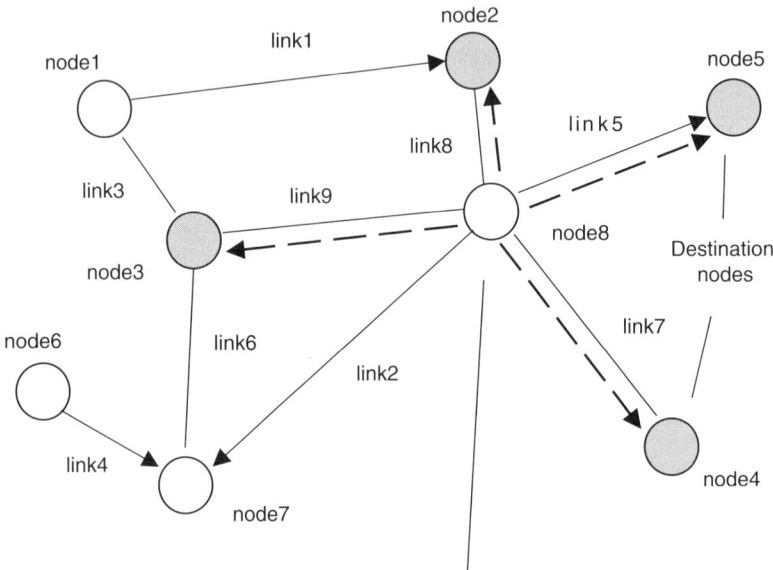

Move via links: link5, link7, link8, link9

Figure 2.9. Local surface broadcasting from a node.

WAVE-WP may be able to penetrate and navigate too. Also the links may be created and maintained that connect its own WAVE-WP nodes and alien nodes, also establishing additional bonds between the nodes all belonging to (same or different) alien systems, which did not exist before the WAVE-WP intrusion.

So the virtual world created by WAVE-WP can be dynamically integrated with other, multiple and dissimilar, systems into a united virtual world in which complex distributed solutions can be effectively found. Its own knowledge network of the WAVE-WP model can also have parts that represent reflection of other systems, including their detailed command-and-control structures, which can be used for an advanced simulation of these systems, ahead of the direct intrusion into their internal organization and behavior.

2.3 UNITED PHYSICAL–VIRTUAL WORLD

2.3.1 The Integration Details

The same WAVE-WP model can work simultaneously with PW and VW, and the distributed world it operates with can be arranged as a combined world in which some parts are physical while other parts are virtual ones. It is also possible to have a much deeper integration of the two worlds within the united physical–virtual world (or PVW, same as VPW) in which any element may have features of both worlds.

Toward this integration, first of all, PW nodes can be connected with other PW or VW nodes by virtual links with certain names (contents). Traversing these links, PW nodes can access (as well as be accessed from) other PW and VW nodes in a surface mode, like in pure VW. As PW nodes are temporary and cease to exist as soon as occupancies in them terminate, any links associated with them will be temporary too.

The PW and VW nodes can also be fully integrated within the united PVW (or VPW) nodes, representing both worlds at the same time. These nodes will be persistent, inheriting this feature from pure VW nodes. They will have both virtual contents (names) and physical coordinates, also an access to physical qualities of the corresponding PW locations, like pure PW nodes do. These nodes will be stationary in the physical world, prohibited from movement in it (contrary to pure VW nodes). Changing their positions in PW, saving individuality and associated occupancies can be possible only by redefining their physical coordinates explicitly.

2.3.2 Access to Nodes in the United World

The united PVW nodes may be accessed directly, in a tunnel mode, by their names or physical coordinates, or both—as having certain names and lying in a proper PW region. They can also be accessed in a surface mode, through virtual links, with additional clarification of the destinations by names or (and) coordinates, as for the tunnel hops. The PVW (VPW) nodes, similar to VW nodes, may also belong to other systems, existing outside the WAVE-WP model.

Different types of nodes (i.e., PW, VW, and PVW or VPW) interconnected by oriented or nonoriented links, both inside and between their classes, may form an advanced representation of the integrated physical–virtual worlds, as shown in Figure 2.10.

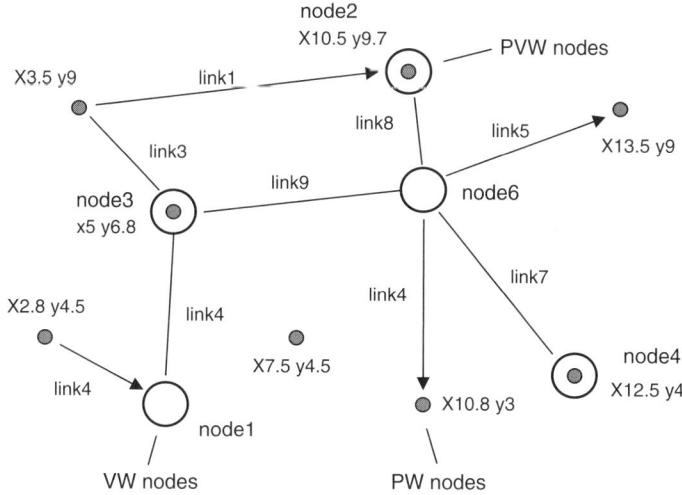

Figure 2.10. Interlinking physical, virtual, and combined nodes within the united world.

2.3.3 United World Dynamics

This integrated and distributed world network model may be highly dynamic, where different types of nodes may appear, change, disappear, or convert into each other's types at run time, involving links between nodes too. As nodes may also change their PW positions at any time, this network may modify, spread, and move (partially or as a whole) in physical space, both changing and preserving its structure.

Such dynamic networks, integrating knowledge of both worlds, can be efficiently used in the creation, simulation, and management of distributed systems of most different natures. The management and simulation of such systems may need moving in physical world first, for performing physical operations and collecting raw data there, from which higher level knowledge may be extracted and processed, enriching the virtual world. The accumulated results in the VW may subsequently or simultaneously be used for advancement in the PW, further obtaining and converting new raw data into knowledge again, and so on.

2.3.4 Time and Speed

Movement in the united world, while forming new PW-bound nodes or changing physical locations of the existing PVW nodes, may need to take into account the notion and value of *time*.

For example, any definition of movement, especially in PW, may be accompanied by setting up a time frame within which it should be completed. Advancing in time may also be needed for simulation and control purposes, where certain operations should start only at a proper moment in time or should be delayed for a certain amount of time. These operations may take place while staying in nodes and doing jobs there or may relate to propagation between nodes, which can be made, as discussed, in a tunnel or surface mode.

The notion of *speed*, similar to time, may also be useful in different operations in the physical, virtual, or united world, for both management and simulation purposes. For example, a recommended or threshold value of speed may be associated with propagation between physical nodes, possibly hinting at proper transportation means for the implementation of the move.

Speed may relate to propagation in PW, for example, in kilometers per hour, or to movement in VW through electronic (cable or wireless) channels, expressed, say, by bits per second. The speed may be set up by its absolute value or by both increment and decrement values—if to be updated during the move in a certain continuum.

2.4 EXECUTION WORLD

The execution world (or EW) is usually hidden from a user of the model, but knowing its basic organization and influencing key parameters may help optimize spatial solutions represented in WAVE-WP.

2.4.1 Doers and Their Connections

All activities in space and time are considered to be performed by *doers*, which may be of any technical or biological origin (humans included). A doer (or a group of them behaving as a separate unit with, possibly, a selected group's representative) can be formally considered as an EW node in WAVE-WP.

These EW nodes may have physical links with each other (say, tactile, vision, sonic, cable, or wireless), which can vary in time when, for example, doers move in physical space and change absolute and relative positions to each other. An example of EW as a dynamic network of nodes formally representing (hardware or software) doers is shown in Figure 2.11.

Staying in EW nodes, it is possible to force doers to move in PW and perform processing of both physical matter (physical objects) and information; these may also be exchanged with other doers (matter included). It may also be possible to set up new physical links with other doers or terminate the existing ones.

Movement of doers in PW may be automatically triggered by the necessity to create new PW and PVW (VPW) nodes, bound to certain locations in physical space, with subsequent PW operations in them. Doer movement may also be necessary to support physical movement of the already existing PVW nodes, or by assisting other doers in performing similar actions. EW nodes (i.e., doers or their groups) can also move in PW on their own initiative, performing jobs scheduled for them individually.

The EW nodes, similar to PW or PVW (VPW) nodes, have coordinates in the physical world that are constantly changing when doers move. EW nodes, if needed, can also be permanently connected, or "pinned," to certain PW locations, becoming

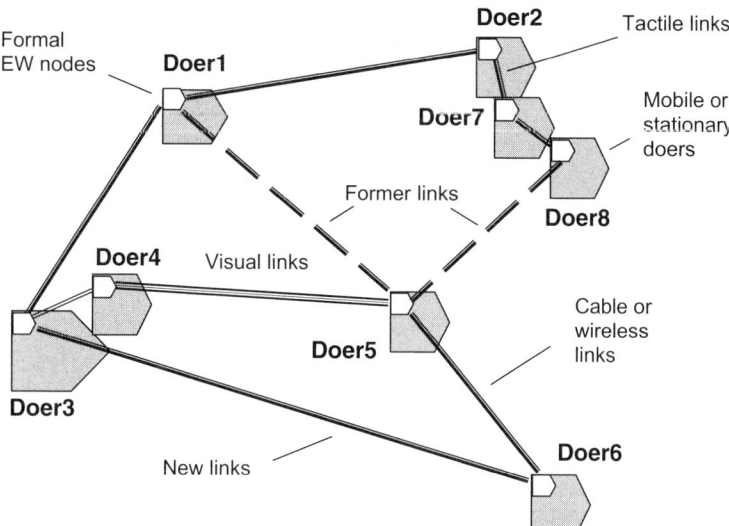

Figure 2.11. Execution world as a dynamic network of doers.

stationary (and prohibited from physical movement). This location dependence may remain indefinitely, unless lifted explicitly.

The EW nodes may have specific names (being the doer names), and physical links between them may hold data classifying their type and capacity. This information can be accessed similarly to how nodes and links of KN are being accessed, as discussed earlier.

2.4.2 Distribution of Physical–Virtual World Between Doers

Doers support subsets of nodes that happened to be located in them with links between these nodes, along with the links to similar subsets in other doers (i.e., leading to nodes in them), as shown in Figure 2.12, with nodes located in doers being named (just numbered).

The PW, VW, and PVW (VPW) nodes may be doer-free or doer-bound (the doer-bound nodes are symbolically connected to the corresponding EW nodes with dotted links in Fig. 2.12). Doer-free nodes can use any available or suitable doer and can move freely between doers (say, due to overload of doers, or when the doers fail to comply with the demands to functionality, speed of movement, or location in PW).

This movement of nodes between doers, which always preserves existing links between them regardless of where they happened to reside at any moment in time, can also be planned and set up explicitly. Doer-bound nodes will have to remain in particular doers unless the change of a doer is set up for them explicitly.

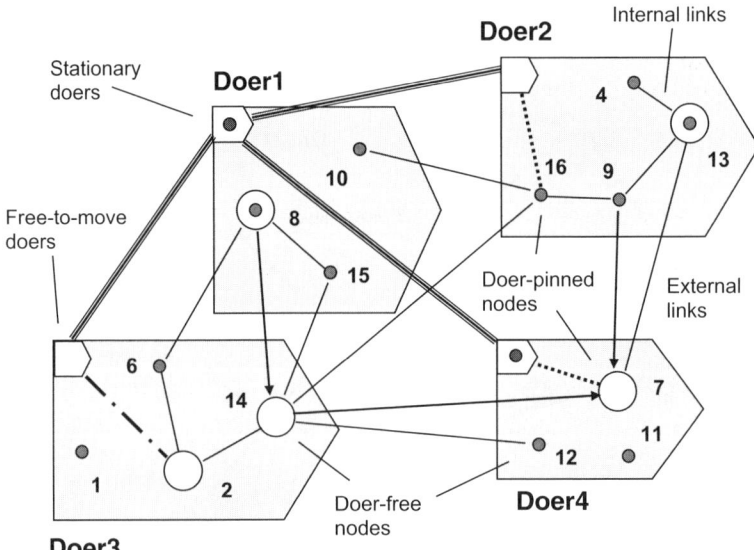

Figure 2.12. Distribution of the united physical–virtual world in the execution world.

When staying in a doer node, it is possible to lift, analyze, and influence basic parameters of the current doer and its physical links with other doers, which can be used in higher level system optimization decisions. One of these decisions may be to move some or all nodes, with associated activities and data, from the current doer to another doer (or doers).

2.4.3 Absolute and Mapping Addresses

The EW nodes have unique *absolute addresses* in the linked EW world. Any other nodes, supported by doers, are supplied in the EW with *mapping addresses*, part of which is an absolute address of the doer that holds these nodes, and another part is a local address (absolute or relative, depending on the implementation) within this doer.

Using these addresses, if they are known (say, copied and recorded in advance), it is possible to directly access the needed nodes in the most efficient way throughout the whole distributed world. The addresses for nodes other than EW ones may, however, not identify the nodes uniquely over time if these nodes can move between doers, thus changing their mappings onto the EW.

2.4.4 Further Integration of Physical–Virtual–Execution World

The EW nodes can also be individually interconnected with other EW nodes, as well as with any PW, VW, and PVW (VPW) nodes (and not necessarily the ones the corresponding doer hosts), by virtual links with any contents and orientations.

Due to such interlinking, these doer nodes can now access (and can themselves be accessed from) any other nodes in the surface mode too. This further integration can help build a really united PW–VW–EW world in which solutions of any distributed problems, including optimization on any levels (the latter including the implementation one), with self-analysis and self-recovery after failures and damages, can be effectively carried out within the same formalism.

2.5 WAVES

2.5.1 Nature of Waves

The WAVE-WP model substitutes the traditional concept of memory, over which computations are usually performed, by the united physical–virtual–execution world, described in the previous sections. This world may comprise a multitude of physical matter, physical objects, information, other hardware or software systems, as well as humans. All these may be complexly interlinked, logically as well as physically.

The world may be active, self-organized, and already controlled by other systems implemented in WAVE-WP or in any other, "alien," model and/or technology. It may also be dynamic, open, and self-evolving in time and space.

Solutions of any problems in this united world are represented in WAVE-WP as its parallel and coordinated *navigation* (or *exploration, invasion, grasping, coverage, flooding, conquest*, etc.) by higher level and ubiquitous powers or *waves*. These can bring local operations, control, and transitional data directly into the needed points (nodes) of the world to be managed or *ruled*.

Waves spread in space and time, providing (multiple) results in the final points (nodes) reached. These nodes and results in them may be returned to a symbolic user as the final solutions or may otherwise be the sources of new waves, which, using data obtained by the previous waves, may invade further parts of the world, obtaining new multiple results in new multiple nodes, and so on. In a most general and abstract form this parallel spatial process is shown in Figure 2.13, with spatial equivalents of such known mechanisms as *branching, looping, intersection, succession, synchronization*, and the like.

The WAVE-WP model allows us to create highly integral, seamless, "agentless" distributed knowledge processing algorithms, internally supported by a variety of special distributed transference, echoing, and command-and-control mechanisms. Some of these mechanisms—such as spatial sequencing and synchronization of parallel actions, spatial looping over distributed domains, and mutual spatial coordination of branches—may be considered as a natural spatial evolution and generalization of the traditional sequential programming over data stored in computer memory.

During this navigation in distributed worlds, which may be loose and free or strictly and hierarchically controlled, waves can modify and restructure the worlds, as well as create them from scratch. Waves may also settle persistent cooperative processes in different points of the worlds, which will be indefinitely governing their further activity and evolution in the way required.

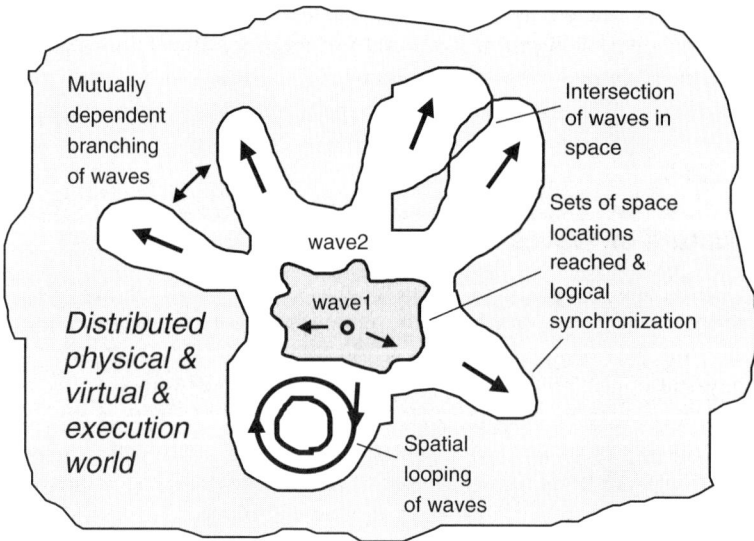

Figure 2.13. Parallel conquest of distributed worlds with waves.

During the distributed navigation, waves create a variety of persistent and temporary infrastructures interlinking knowledge, processes, and computers. This allows the distributed solutions to be highly integral ("conscious"), goal-driven, internally and externally controllable, and self-recovering after damages.

2.5.2 Navigation in Space

Waves, starting from certain points (nodes), navigate in space. During the navigation, which may be parallel and under speed and time constraints, waves may stay indefinitely in nodes and perform local operations in them; they may form new links and nodes with moving, correspondingly, via and into them; or they may enter the already existing nodes, formed earlier by the same or other waves.

Waves may also create and leave additional links between the already existing nodes, which reflect new relations between them. So navigating in the worlds, waves may create these worlds too, possibly, starting from an empty space, that is, from scratch.

Moving in some direction in space can be influenced by the success or failure of moving in other directions, and vice versa, thus forming dynamic spatial-logical coordination and control between different wave branches. This coordination may be hierarchical and recursive.

Navigation of the world may start from many points (and by many users), simultaneously, and the spreading fields of solutions, which originated from these points, may overlap and cooperate (or compete) in the distributed space and time continuum or can be independent from each other.

Waves may be symbolically considered as progressing in parallel between different sets of nodes reached (or SNR) in space (with initial SNR containing the starting node only), where transitions between adjacent SNRs may be gradual and asynchronous, rather than globally leveled and synchronized. Subsequent SNRs may contain nodes from any previous SNRs, thus reflecting distributed looping in space. Same nodes may be represented more than once in an SNR, when, say, waves split in nodes or their different branches enter the same nodes from other nodes, so SNRs may not be classical sets in general.

2.5.3 Actions in Nodes

Arbitrary actions can be performed while staying in nodes, including accessing and changing physical world parameters for PW-related nodes, updating the node and (passed) link contents, removing the nodes together with all adjacent links, as well as creating new nodes and links. Local results obtained in nodes may be left there and shared with other waves entering these nodes.

Other data can navigate with waves further, as their sole property, supplying subsequent navigation stages (and the nodes reached) with the results obtained in the previous stages. The data staying in nodes or moving with waves may represent *information, physical matter*, or any *combination* thereof.

Staying and acting in nodes, waves may search physical or virtual as well as execution space to any depth by launching subordinate waves. The latter may collect, process, and return remote data for further processing in the current nodes or may change data in remote nodes. Waves may also force nodes move in both PW and EW, while preserving links between themselves and with other nodes. This node movement in space can also involve all waves and temporary data currently associated with these nodes, which will accompany the nodes as their property.

2.5.4 Coverage with Rules

Multiple waves, describing simultaneous navigation and processing in space, may be covered by distributed control, expressed by means of special nonlocal coordination constructs, called *rules*. Rules manage cooperative, competitive, or independent conquest of space in both breadth and depth mode and may add certain general functionality to waves. They may, for example, set up a special creative mode over the spreading waves, where new nodes and links will be formed during the space navigation, or can detail how the obtained distributed results should be collected and returned for a further processing.

The generality of rules also allows WAVE-WP to be effectively used as a traditional (sequential or parallel) universal programming language, so there may be no need to use other languages for a variety of application purposes.

2.5.5 Composition and Structuring of Waves

The WAVE-WP model universally characterizes any movement or action in space, whether simple or complex, by the following three features:

1. Set of nodes reached, which may be remote
2. Combined value obtained in these nodes, which may be multiple
3. Resultant generalized termination state

Taking these three into account, any space-conquering and knowledge-processing problems can be expressed by the model, often in a very simple and dense form, with an effective and convenient composition of space-navigating waves.

We will consider only elementary examples of composition of waves, illustrating how they navigate in the distributed world, which may be of any PW, VW, or EW origin, or their integration. In the following examples, w1 to w9 may be arbitrary complex waves.

In a *sequential composition* of waves (separated by a semicolon), each new wave starts from all positions in space reached by the previous wave (i.e., its SNR), as shown in Figure 2.14 for the sequence of waves w1 and w2.

Starting in the node Start and applying w1 in it, SNR1 will be obtained somewhere in a distributed world. Then to each node of SNR1 the same wave w2 will be applied (by being replicated), which will altogether produce SNR2 in space—the

WAVES 51

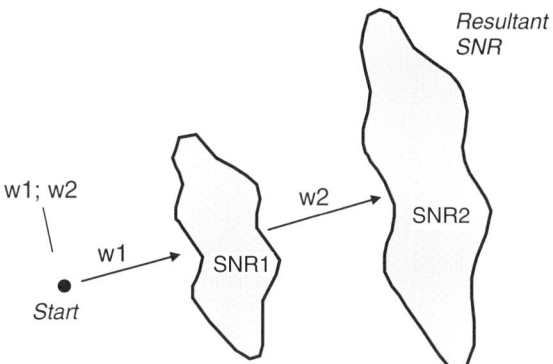

Figure 2.14. Sequential composition of waves.

latter being the final set of locations reached in space on the sequence of these two waves.

Grouping the reached nodes by w1 and w2, correspondingly, into SNR1 and SNR2 is purely symbolic and conceptual, as w2 may start developing from the nodes reached by w1 immediately, without waiting for the whole SNR1 to be completed, and the used nodes of SNR1 may start disappearing as soon they have been abandoned by w2. Also, the individual nodes of SNR2 may cease to exist immediately after w2 terminates in them, not waiting for the whole SNR2 to be completed. All these navigation processes may thus be fully asynchronous and parallel in space.

In a *parallel composition*, all waves (separated by comma) start developing from the same nodes independently and in parallel, forming altogether the resultant SNR that may not exist as a whole at any moment in time too, with nodes, say PW ones, terminating as soon as waves in them terminate. The development of parallel composition of waves w1 and w2 is shown in Figure 2.15.

In a combined *parallel–sequential composition* of waves, every group of parallel waves forms altogether a united SNR, from every node of which another parallel group develops independently, replicating as a whole for each node it starts from. The final SNR on this composition will correspond to all nodes reached in space by all instances of the latest parallel group in the sequence. This is shown in Figure 2.16 for the composition: w1, w2; w3, w4, representing a sequence of two parallel groups, each one having two independent waves.

Using rules For more complex cases of parallel–sequential structuring, parentheses can be used, as well as rules, the latter adding proper constraints or contexts to the development of waves. For example, in the following composition:

or(w1, w2, w3); w4

the rule or will activate from the same node three waves w1, w2, and w3 in a sequence and will stop with the first wave that terminates with a nonempty SNR. This SNR will be declared the resultant SNR on the whole group of the three waves, and from this

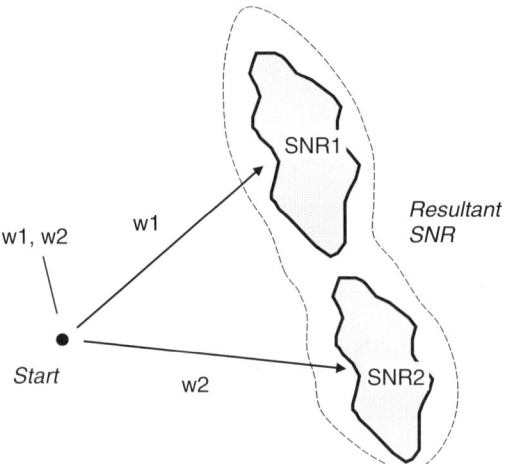

Figure 2.15. Parallel composition of waves.

SNR (from all its nodes) the wave w4 will be developing in a replicated mode, producing by all its instances the final SNR of this composition of waves.

The work in space of this ruled composition is shown in Figure 2.17, where wave w1 is supposed to fail from the start node (producing empty SNR), and wave w2

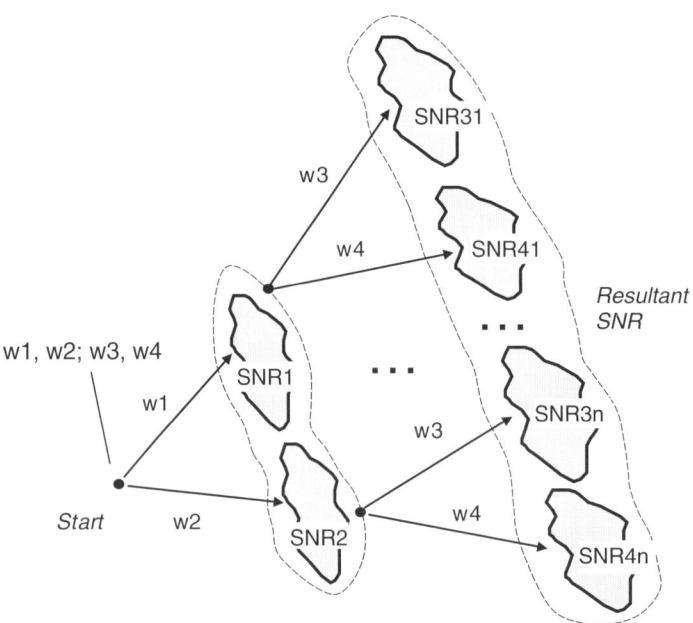

Figure 2.16. Parallel–sequential composition of waves.

WAVES

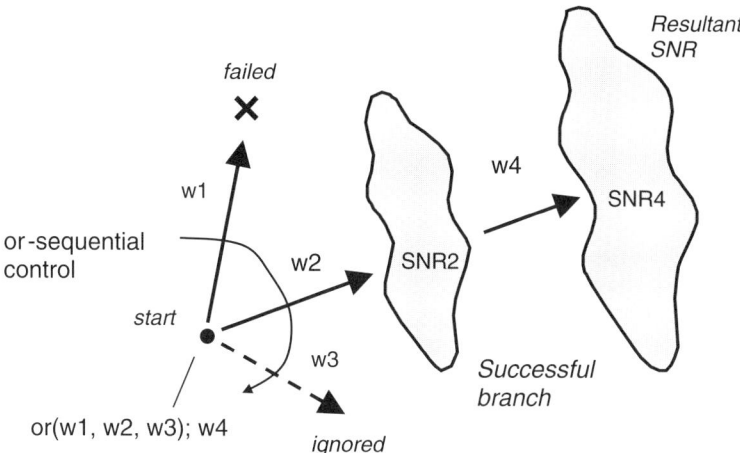

Figure 2.17. Example of using or-sequential rule.

appeared to be the first successful wave in the group; from its SNR the rest of the composition, that is, w4, will be applied.

2.5.6 Wave Expressions and Remote Data

The WAVE-WP model, contrary to its predecessor WAVE, allows us to have arbitrary complex expressions operating with data that may reside everywhere in the distributed world, with parts of the expressions providing an access to this data, while other parts setting operations on the (remote) data returned by this access. We will show here some very simple examples of these possibilities.

Assigning Remote Results to Local Variables. First, let us see how we can directly assign multiple results obtained remotely to local variables when staying in some point (node) of the distributed united world, using for this, for example, the following expression:

```
Nresult = (andparallel(w1, w2); w3)
```

This will assign the final results that can be reached in space by parallel waves w1 and w2, followed by w3, to the local variable Nresult in the node where this program starts. Only success of both w1 and w2 (i.e., nonempty SNR of each of them), which is coordinated and checked by andparallel rule, will allow wave w3 to be applied simultaneously in all nodes of SNR1 and SNR2, in a replicated mode.

This will lead to the final SNR of the right hand of the assignment. And the final values obtained in all nodes of this final SNR (which may be scattered throughout the whole world) will be collected and assigned together as a new content of Nresult in the starting node. These spatial processes are illustrated in Figure 2.18.

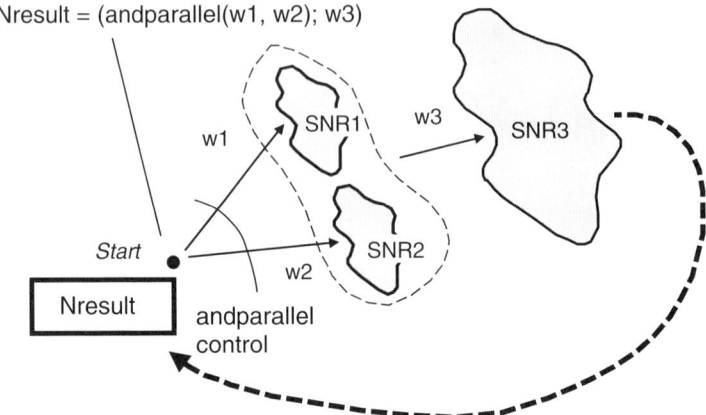

Figure 2.18. Assigning remote results to a local variable.

Assigning Local Results to Remote Variables. Let us consider a reverse example, where local results are assigned to remote variables, using for this the following expression:

(w4; w5; N12, N66) = (3.4, 6.9)

At the left hand of the assignment, first wave w4 will develop from the node to which this expression is applied, and then from each node of its SNR wave w5 will be replicated and applied, leading to the final SNR on these two waves. In each of these finally reached nodes, local variables named N12 and N66 will be assigned the new contents as a vector given at the right hand of the assignment. If such variables do not exist in these final nodes at the left, they will be created there and loaded with the right-hand vector value. These spatial processes are shown in Figure 2.19.

Assigning Remote Results to Remote Variables. Arbitrary complex expressions operating directly with any data distributed throughout the united world can be composed and executed within the WAVE-WP model. For example, it is trivial to combine the previous two cases within a single expression that will be assigning multiple remote results, obtained in space when starting from some node, to multiple remote variables, reached from this node, as follows:

(w4; w5; N12, N66) = (andparallel(w1, w2); w3)

The work of this combined expression is shown in Figure 2.20, where remote results obtained at the right hand of the assignment will be directly transmitted and assigned to remote variables reached (or created) at the left hand. The node from which this expression has started will be serving as the logical (as well as "hourglass") center of the whole distributed scenario.

The waves at the left and right parts of the assignment may develop in parallel from the very beginning. The right part will be propagating in space for obtaining its final

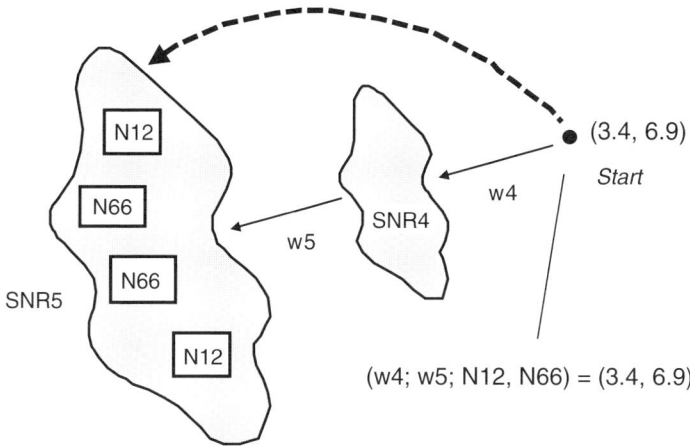

Figure 2.19. Assigning local results to remote variables.

results, and the left part will be moving (possibly, by quite different directions in space) to find existing (or create new) variables with proper names and locations. To these remote variables, when they are prepared, the remote results from the right, when ready too, will be directly transmitted.

2.5.7 Delivery and Processing of Physical Matter

The values reached and (or) obtained in space and assigned to variables in WAVE-WP may not only be information but physical matter (or physical objects) too. So the following expression, reaching in space and picking up the needed amounts of

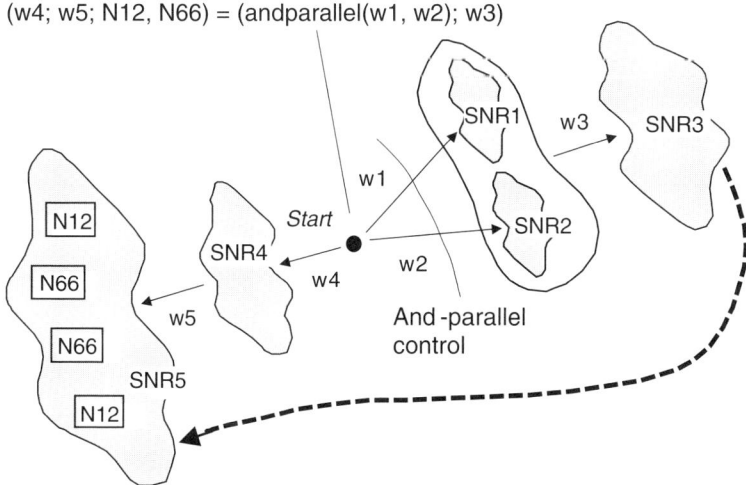

Figure 2.20. Direct assignment of remote results to remote variables.

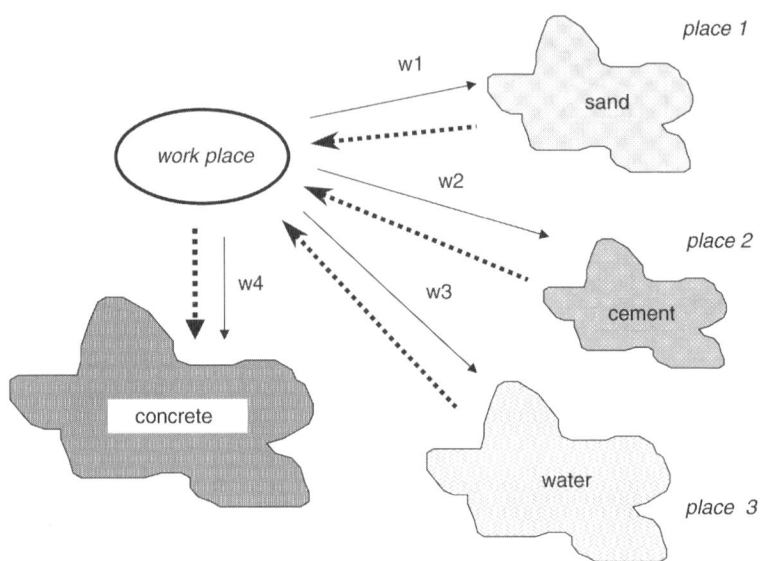

Figure 2.21. Direct operations on a remote physical matter.

sand, cement, and water (correspondingly, by waves w1, w2, and w3), with a subsequent collecting and adding them to each other to produce concrete and store it in a remote variable Nconcrete (reached by wave w4), will be legal in WAVE-WP.

```
(W4; Nconcrete) = (w1; "3 tons of sand") + (w2; "2 tons of cement")
+ (w3; "4 tons of water")
```

Physical matter items, to distinguish them from information, are enclosed in double quotation marks. The node in which this expression starts will be considered as an actual or symbolic (depending on the implementation) working place for mixing the returned physical ingredients and producing the resultant matter (concrete), before delivering the latter to the proper destination, using wave w4. The parallel development of waves and movement of physical matter in PW for this example are depicted in Figure 2.21.

2.6 CONCLUSIONS

We have considered main features of the WAVE-WP model that can directly process and control different types of distributed worlds, namely continuous physical, discrete and networked virtual, and dynamically interconnected world of computerized doers of any technical or biological origin, as well as any combinations of these.

CONCLUSIONS

Spatial integration and mapping of these worlds onto each other may be dynamic and controlled, allowing application and internal system optimization problems to be solved coherently within the same space-conquering formalism. WAVE-WP also represents a higher level distributed processing and control model than its predecessor WAVE, allowing us to compose arbitrary space-processing expressions operating directly with remote data, the latter representing both information and physical matter or physical objects.

We have given only a general, informal description of the WAVE-WP model, which will be further clarified by a detailed description of WAVE-WP language in the following chapter.

3

WORLD PROCESSING LANGUAGE

The WAVE-WP language operating with the united physical–virtual–execution world discussed in the previous chapter will be described. Its peculiar syntax and semantics allow us to effectively oversee distributed worlds and systems of different natures and establish any needed control over them.

It is a higher level language for spatial processing and control, which can *directly or inferably* access *any* points of distributed worlds, *directly* process *any* local or remote data, as well as *directly* assign *any* local or remote results to *any* local or remote variables. The language operates on values that may represent information, physical matter (physical objects), or any combination of the two.

A run time creation and modification of physical, virtual, and execution worlds, as well as change of mappings between these worlds within the united world, can be efficiently expressed in the language.

Detailed syntax and semantics of WAVE-WP will be presented along with elementary programming examples explaining them. Due to the new world-conquering philosophy underlying the language and also numerous extensions, changes, and improvements in both syntax and semantics as regards the previous language versions, WAVE-WP will be described here as a new language and not as an extension of WAVE published in a previous book (Sapaty, 1999a).

Ruling Distributed Dynamic Worlds, by Peter S. Sapaty
ISBN 0-471-65575-9 Copyright © 2005 John Wiley & Sons, Inc.

3.1 TOP LANGUAGE ORGANIZATION

The WAVE-WP language has simple recursive syntax, with its top level expressed as follows:

$$
\begin{aligned}
wave &\longrightarrow \{\, advance\,;\,\} \\
advance &\longrightarrow \{\, move\,,\,\} \\
move &\longrightarrow constant \mid variable \mid \\
&\qquad \{\, move\ act\,\} \mid \\
&\qquad [\,rule\,]\,(\,wave\,)
\end{aligned}
$$

In this description, words in italics represent syntactic categories, braces show zero or more repetitions of the construct, with a delimiter at the right if more than one construct in a sequence, the delimiters being semicolon, comma, and acts symbols. The vertical bar separates alternatives, and square brackets identify an optional construct. Semicolon and comma as the language symbols separate, respectively, successive and potentially parallel parts. Parentheses in the language are used for structuring of waves.

A program in the language is called a *wave*. Its successive parts, or *advances*, develop from all nodes of the set of nodes reached (or SNR) of the previous advance. Parallel parts, or *moves*, belonging to the current advance, develop from each node of the SNR of the previous advance independently (in any order or in parallel if not stated otherwise by rules of the language), adding their own SNRs to the whole SNR of the current advance.

Moves can be of the following types:

- First, they can point at a resulting value, as a *constant* or *variable*, directly.
- Second, they can be any data-processing and space-navigating *expressions* consisting of parts (moves again, due to recursion) separated by elementary operations (*acts*). Acts are working on values (which may be local or remote) provided on the left and right sides of them as their *operands*.
- Third, they can themselves be arbitrary *waves in parentheses*, optionally prefixed by *rules*. If such moves are part of an expression, bound by acts to the rest of it (as in the previous case), they are lifting results obtained in their final SNR and returning them to the expression that activated them.
- Otherwise, if standing independently in an advance (possibly, separated by commas from other moves), they are delivering the program control into their resultant SNR, from which the next advance will be applied.

All acts in an expression are executed in the node they start, requesting to this node the data from their left and right operands (which can be waves, due to recursion) and leaving the obtained results in this node too, to be used by other acts. After termination of the expression, the program control will remain in the current node or will shift to other nodes if a hop was used as the last act.

TOP LANGUAGE ORGANIZATION   61

EXAMPLES OF STRUCTURING OF WAVES. In the following examples, w1 to w6 may be arbitrary complex wave programs (if needed, in parentheses to obey the language syntax).

- Sequential composition:

```
w1; w2; w3
```

The three waves, as consecutive advances, starting in the current node will develop in physical or virtual (as well as combined) space one after the other, with each new wave starting from the SNR formed by the previous wave. No synchronization of these distributed processes will take place, that is, each advance may start developing from emerging SNR nodes of the previous advance immediately, without waiting for the completion of this SNR.

- Independent or parallel composition:

```
w1, w2, w3
```

The three waves, as parallel moves of a single advance, all starting in the same current node, will develop independently in space, with SNR of each of them subsequently merging into a united SNR on the whole advance (here the whole wave). No synchronization will take place.

- Parallel–sequential composition:

```
w1, w2, w3; w4, w5, w6
```

The wave has two consecutive advances, each comprising three waves as moves, independent of each other. The second advance will develop from all nodes of SNR of the first advance, this SNR uniting SNRs obtained independently on waves w1, w2, and w3. The resultant SNR on the wave will comprise SNRs formed independently by waves w4, w5, and w6.

- Data processing expression:

```
Name = w1 + w2
```

Here waves w1 and w2 are supposed to result with numerical values in their SNR, which may be remote; these values (generally as vectors) will be brought to the current node and added to each other, with result assigned to nodal variable Name.

- Structuring with the use of rule:

```
sequence(w1, w2); w3
```

Here `sequence` is one of rules of the language. Wave `w1` will develop in space first, and only after its full termination `w2` will start. Wave `w3` will begin from all nodes reached by both `w1` and `w2`.

3.2 DATA DEFINITIONS

3.2.1 General on Constants

There are no type definitions in the language, and any data is self-identifiable by the way it is written in a program. Data constants may be of the following categories:

$$constant \longrightarrow information \mid physical\text{-}matter \mid special\text{-}constant \mid$$
$$aggregate\text{-}constant$$

Information. Representation of information may have the following two options:

$$information \longrightarrow [\ sign\ ]\ \text{'}\{\ character\ \}\text{'} \mid [\ sign\ ]\ \{\ \{\ character\ \}\ \}$$

Strings of characters in single opening–closing quotation marks stand for a most general representation of information. The quotation marks may be nested, appearing inside the string too. Arbitrary information may be represented in such a way, including program code in WAVE-WP or any other language, to be executed.

Strings of characters that can be potentially used as program code in WAVE-WP can also be put in braces. This will automatically trigger, when accessed, their parsing and optimization as a potential WAVE-WP language program, which may speed up their future execution. Braces may be nested in the string too.

Information strings can be prefixed by a sign for special cases, like setting both link orientations and their contents when moving through existing knowledge networks or creating their new parts in hop operations, as explained later. Beyond these special cases, the sign preceding quoted or braced strings has no effect.

Physical Matter. For representing physical matter or physical objects in general, double quotes are used:

$$physical\text{-}matter \longrightarrow \text{"}\{\ character\ \}\text{"}$$

It is assumed that contents of such double-quoted strings are recognized by the system implementation, which has embedded capabilities for direct processing of certain physical matter or physical objects.

Special Constants. Single and double quotes, as well as braces, may be omitted for special constants, which are self-identifiable by the way they are written, while

DATA DEFINITIONS

strictly obeying the language syntax. These special constants are listed and explained in the following section.

Aggregate Constants. Different elementary constants, either general or special, can be linked together by underscore to form aggregate constants, reflecting joint (often multiple-type) values, which can be treated as a single data unit in special operations, as follows:

$$aggregate\text{-}constant \longrightarrow \{\ constant\ \_\ \}$$

3.2.2 Special Constants

These constants may be used without quotes or braces, as already mentioned, and may be of the following types:

$$special\text{-}constant \longrightarrow number \mid place \mid address \mid time \mid speed \mid node\text{-}kind \mid$$
$$doer \mid infra\text{-}link \mid state \mid auxiliary$$

Their syntax and brief semantics are revealed below, but details of use being explained later throughout this chapter, when acts and special variables are considered. It is also possible, albeit not recommended, to use other quoteless constants, which may not be classified as the special constants described below and which do not confuse the language syntax. These constants will be treated as general strings.

Numbers. Numbers follow the traditional syntax, which covers both integer and real values, as follows:

$$number \longrightarrow [\ sign\ ]\ integer\ [\ .\ integer\ ]\ [\ e\ [\ sign\ ]\ integer\ ]$$

where *integer* is a sequence of *digits*, and *sign* has two values, as follows:

$$integer \longrightarrow \{\ digit\ \}$$
$$sign \longrightarrow +\mid -$$

Places. A place in PW can be represented by a sequence (of identifiers) of *dimensions* followed by *numbers* expressing values of these dimensions:

$$place \longrightarrow \{\ dimension\ number\ \}$$

The place can be a point in PW set up by absolute coordinates or by coordinate shifts from another place, where dimensions x, y, and z are used for identifying absolute coordinates, and dx, dy, and dz stand for their shifts from a certain point.

The place can also represent a region in PW with its *range*, or *radius*, used as an additional dimension to the coordinates of its center as a point (r being the identifier for range). So the full set of dimension identifiers is as follows:

$$dimension \rightarrow x \mid y \mid z \mid dx \mid dy \mid dz \mid r$$

The order of dimensions in representing a place may be arbitrary, and no special delimiter is required between them, as the dimension identifiers serve as delimiters themselves between different values in the place definition. However, for improving readability, free spaces may be added between different dimensions, as they have no value (unless in quotes) for the language syntax.

Addresses. Address of a node in VW is generally represented by a sequence of integers separated by a period, this sequence being prefixed by letter a:

$$address \rightarrow a \ \{ \ integer \ . \ \}$$

The sense of different integers may depend on the implementation, describing, say, the addressing hierarchy in a multilayered distributed system.

Time. Time may also be set up as a sequence of integers (representing, say, years, months, days, hours, and seconds), all separated by a period and prefixed at the start by t for the absolute time or dt for its positive shift from the current moment, as follows:

$$time \rightarrow t \ \{ \ integer \ . \ \} \mid dt \ \{ \ integer \ . \ \}$$

Speed. Speed of movement in the integrated world may be represented by a number prefixed with s for its absolute value or ds for a shift value (incremental or decremental), as follows:

$$speed \rightarrow s \ number \mid ds \ number$$

Its treatment, say, as kilometers or miles per hour or bits per second, may depend on the context of use and system implementation details.

Kinds of Nodes. Node types, or *kinds*, are represented by a sequence of letters p, v, and e defining the belonging and (or) connection of a node, respectively, to physical, virtual, or execution world, with the allowed combinations being as follows:

$$node\text{-}kind \rightarrow p \mid v \mid e \mid pv \mid vp \mid pe \mid ve \mid pve \mid vpe \mid ep$$

Clarification of the use of these types will be explained later in relation to a special variable responsible for the node kind value.

Doers. Doers may be considered as real physical objects or as a reference to them only (i.e., information), depending on the context of use in the program. Doers can be represented by a sequence of alphanumeric, or *alphameric*, values separated by a period (i.e., consisting of letters and/or digits only, with at least two values in the sequence):

$$doer \longrightarrow \{\ alphameric\ .\ \}$$

The doer names used should be distinguishable from other special quoteless constants described in this section. Using single or double quotation marks may give more freedom for naming doers, which, of course, should be recognized by the interpretation system.

Infrastructure Link. This constant reflects a link of the current node to one of temporary internal interpretation infrastructures covering, or passing through, this node. This link is represented by an integer prefixed by i, as follows:

$$infra\text{-}link \longrightarrow \texttt{i}\ integer$$

"Diving" into internal interpretation layer from the application layer may be useful for writing effective spatial programs in certain cases. The use of infrastructure links is explained further in connection to a special infrastructure variable.

States. The state constants, which are used to generate control states in the program explicitly, may have the following five values:

$$state \longrightarrow \texttt{abort}\ |\ \texttt{thru}\ |\ \texttt{out}\ |\ \texttt{done}\ |\ \texttt{fail}$$

Their meanings and use in waves will be explained later.

Auxiliary Constants. There are also a number of auxiliary keywords (some may be prefixed by a sign), formally classified as constants too:

$$\begin{aligned}auxiliary \longrightarrow &\ \texttt{direct}\ [sign]\ \texttt{any}\ |\ \texttt{all}\ |\ \texttt{noback}\ |\ [sign]\ \texttt{infinite}\ |\ \texttt{start}\ |\\ &\ \texttt{back}\ |\ \texttt{fringe}\ |\ \texttt{nil}\end{aligned}$$

where direct, any, all, and forward are used in hops; start, back, and fringe help to detail movement through the existing internal infrastructures; infinite stands for an arbitrary large (positive or negative, if signed properly) value; and nil is for *nothing* (no value or object, i.e., empty). Their usage will be clarified in the context of different operations of the language.

EXAMPLES OF CONSTANTS

- Numbers: 3, 66.5, 1.5e-6
- Address in VW: a124.66.06
- Node kinds: pv, vpe, ep
- Doers: robot.66, "carrier.44"
- Termination states: done, abort, out
- VW node names: 'friends', 'Peter and Paul'

- Wave program: `{N = or('link' # any, 1000)}`
- Physical matter: `"water"`, `"sand"`
- Auxiliary constants: `direct`, `+ any`
- Aggregated constant: `'target2'_x5.6 y77 z88.5 r3.1_dt66.2`

3.2.3 Vectors

Different acts of the language may be performed not only on single constants (elementary or aggregate, as discussed above) but also on their ordered sequences symbolically called *vectors*, while modifying existing or producing new vectors. Vectors may also be recorded as the contents of variables.

Vectors in WAVE-WP are formally wave programs too, expressed in the language syntax already discussed. They consist of values as parallel moves separated by commas and are enclosed in parentheses. The number and types of their elements are not declared in advance, and these may vary at run time. Different acts may operate on vectors as a whole or on their separate elements.

Elements, or *scalars*, of a vector, which may be of different types within the same vector, can be *indexed*, with the indices starting from one. Separate vector elements may be directly accessed and updated by both their indices and contents (not single elements only but their groups as well, using vectors of indices or contents for addressing multiple elements).

Vectors may be treated as both ordered sequences and sets, with corresponding operations on them. Vectors may, however, have repeating elements thus differing from classical sets.

As `nil` in the language means really nothing, it cannot exist as an element of a vector. If written as a move in a wave, it will be omitted in the evaluated wave if the latter is used as a vector.

EXAMPLES OF VECTORS

- `(5, 6, 44, 4)` — a vector consisting of four elements of the same type.
- `(5, "brick", "water", 'one two three', {5-6}, done)` — a vector consisting of different type elements.
- `N = (66,'alpha' && 'beta', nil,, 88.7, nil, 3, 3 * 5-1)` — the right part of the assignment will be considered as a wave to be executed (evaluated) first. The result of this execution, consisting of constants only, after removal of empty (or `nil`) elements from it, will be assigned to variable N. The resultant vector value associated with variable N will be as follows: `(66, 'alphabeta', 88.7, 3, 14)`.

3.3 VARIABLES

There are three classes of variables with which the waves operate, differing from each other by where they are located, how they are used, and which kind of information they hold or access.

$$variable \longrightarrow nodal \mid frontal \mid environmental$$

Similar to data types, there are no declarations of variables in the language, and their classes are recognized by the way their identifiers are represented. Variables can also be associated with vectors, which may have a changing number of elements. The latter may be of different and changing types too.

3.3.1 Nodal Variables

Nodal variables (with identifiers prefixed by M or N) are created in nodes by waves visiting these nodes and may remain there being shared by different waves traversing the nodes.

$$nodal \longrightarrow M\,[\,alphameric\,] \mid N\,[\,alphameric\,]$$

Nodal variables, being automatically created by the first access to them by some wave in the node, can hold (or can be associated with) both information and physical matter. They are considered stationary in relation to nodes they belong to but can move together with these nodes if the latter change their positions in the united world. Nodal variables cease to exist when the nodes they are associated with are removed.

Variables with identifiers starting with M can be shared by any waves visiting these nodes. They remain attached to these nodes until these nodes exist, or unless being removed explicitly by some wave (assigning nil to them), or by a general system cleanup activity launched periodically, say, when the lifetime of these variables exceeds a certain system threshold.

Variables having identifiers starting with N can be accessed in a node only by waves having the same identity, or *color*, while being invisible to waves with other colors (the waves coloring policy will be explained later). Many variables with the same name can exist in the same nodes, each serving waves with a particular color—the one within which they were created.

The N variables are automatically removed in a self-cleaning mode, when the whole wave program that produced them (or its separate part covered by a special rule) terminates. This rule causes independent termination of the program part, regardless of the state of the rest of the program.

3.3.2 Frontal Variables

Frontal variables (their identifiers starting with F) belong to waves, not nodes, and navigate in space with the waves.

$$frontal \longrightarrow F\,[\,alphameric\,]$$

Frontal variables cannot be shared with other waves (or with other threads of the same wave program). They replicate too when their waves split or replicate, and each new branch will have an independent copy of the variable operating with it individually.

Similar to nodal variables, frontal variables can hold both information and physical matter. Their automatic replication is trivial for information, but for physical matter frontal variables can be copied only if this is supported by the implementation system (which, say, can reproduce physical matter or object as regards the specifics of this matter).

3.3.3 Environmental Variables

Environmental variables (each having a special name) access different properties of the navigated worlds, also time and speed, as well as some internal features of the implementation system. The following identifiers are used for these variables:

$$\textit{environmental} \longrightarrow \texttt{KIND | QUALITY | CONTENT | LINK | ORDER |}$$
$$\texttt{WHERE | TIME | SPEED | DOER | JOINT |}$$
$$\texttt{ADDRESS | INFRA | BACK | VALUE | PAINT |}$$
$$\texttt{RESOURCES | USER | GROUND}$$

Let us consider the use of these variables in wave programs.

KIND. Returns *kind*, or type, of the current node, with its possible values and their use being as follows:

p	PW location as a temporary node
v	Persistent VW node
e	Doer, or EW node
pv	Combined PW-VW (or PVW) persistent node
vp	Combined VW-PW (or VPW) node, same as previous
pe	Doer-bound PW node
ve	Doer-bound VW node
pve	Doer-bound PVW node
vpe	Doer-bound VPW node, same as previous
ep	PW-bound, stationary doer node

By assigning to KIND, it is possible to change types of nodes at run time. The following are some examples.

Changing type from p to pv (or vp) results in converting this temporary PW node into a persistent one, with some default VW name and a certain address in the virtual world. Changing the node's type from v to vp (pv) connects the virtual node to the position in PW where the doer hosting this node currently stands, making the node stationary in the physical world.

Changing the type from pv (vp) to v removes dependence of this node from PW, making it free to migrate in physical space. Changing pe, ve, and pve (or vpe) to, correspondingly, p, v, and pv (or vp) releases these nodes from their dependence from the EW world, permitting them to migrate freely between doers at run time. The opposite change will pin these nodes to the doer they are currently staying in, and this binding will remain indefinitely, unless changed explicitly.

In a more complex example, changing v to pve (or vpe) will simultaneously bind the virtual node to its current doer and the physical location this doer currently stands in, whereas for the reverse change, the node will become both PW and EW free.

Changing from e to ep will pin the current doer to its current physical location, making this doer stationary in PW, along with all nodes it currently hosts. Such a PW pinning of nodes via the doer that hosts them is, however, not absolute, as they can potentially change the doer and become PW free (if are not associated with PW locations directly).

QUALITY. If a node is of p, pv (vp), pe, or pve (vpe) types, QUALITY accesses in this PW location a vector of physical parameters chosen to represent this world within the particular implementation (say: pressure, temperature, density, humidity, radiation level, visibility, etc.). By assigning to QUALITY in these nodes, a certain impact on the PW can be made, where the whole vector of physical parameters, or a group of them, can be influenced simultaneously. Accessing this variable in v-type nodes will link to the current doer's location in PW (same as if standing in the corresponding e node itself).

Changing PW by just assigning to QUALITY in its different points may, however, be problematic in general and may depend on a concrete system implementation and its vision and treatment of the environment.

CONTENT. Accesses the content, or name, associated with the current node, if the node has its presence in VW (i.e., being of v, pv, vp, pve, or vpe types) or is a doer (i.e., of e or ep types). CONTENT may hold an arbitrary string scalar value, including a program in WAVE-WP or any other languages.

Assigning to CONTENT in v, vp (pv), or vpe (pve) nodes changes the node's content. If a vector with more than one element is trying to be assigned to CONTENT, only its first scalar will be used, while others skipped. Assigning nil to CONTENT in these VW-bound nodes results in removal of the current node, together with all temporary information (kept in nodal variables) associated with it, as well as deletion of all its links to other nodes. This removal is *graceful*, waiting until after all waves in the current node terminate (similar to the termination of temporary PW nodes).

Assigning nil to CONTENT in a doer node may be used for a user-controlled shutting down of this doer, after causing all nodes it hosts (together with local information and current waves in them) to move to other doers, which should be carried out automatically by the interpretation system. So this node's symbolic removal is planned to be graceful too. If such transference of the load to others is not possible, the operation will fail.

If the node is of p or pe types, CONTENT returns nil (as pure PW nodes do not have identity), and assigning to CONTENT causes no effect.

LINK. Returns content (name) of the link that was passed when entering the current node, which may be of any scalar value, the same as the node's content. The passed link's content may be changed by assigning any value to LINK. Removing the passed link can be done by assigning nil to LINK. Similar to nodes, assignment of only a scalar value to LINK is possible. If the current node has been reached in a direct (tunnel) hop, LINK returns nil if accessed, and any assignment to it causes no effect.

ORDER. Returns orientation, or order, of the passed link reflecting how it was passed: + if the link was passed along orientation; - if opposite the orientation; and nil if the passed link was nonoriented. Assigning these values to ORDER may change the orientation of the link. If the current node has been reached in a tunnel hop, ORDER always returns nil, with any assignment to it, similar to the assignment to LINK, causing no effect.

WHERE. Returns for PW-bound nodes (i.e., of p, pv, pe, pve, or vpe types) coordinates of the location in PW assigned to the current node (which may differ from the current doer's PW position, as the doer may serve other nodes). If the node is of ve type, WHERE returns current coordinates of its current doer. The same takes place when standing in the doer (e or ep) node itself. Standing in a v-type node, WHERE always returns nil.

Assigning to WHERE in a PW-bound node changes the physical position associated with this node and may cause the current doer to move to this new location. Assigning to WHERE in e or ep nodes orders the doer to move to the new location immediately. Any assignment to this variable in v-type nodes causes no effect. Staying in any node, it is always possible to know its current actual position in PW by first hopping into its doer node (see later), and then accessing WHERE in it.

Changing coordinates by assignment to WHERE may be done by both their absolute and shift values.

TIME. Returns the current global time value. Assignment to TIME, however, does not change global time but rather suspends activity of the current wave in the current node until the new time value equals the growing global time. The value assigned to TIME must surpass the current global time. Assignment to TIME may be by both its absolute and shift value.

SPEED. Returns the speed value associated with the current PW-bound node in the physical world (which may differ from the current speed of the current doer, if this node is not the doer itself). Assignment to SPEED changes the recorded node's speed and, in case of doer node, updates its actual speed directly. Knowing the actual current speed of any node is always possible by hopping to its doer node, and then measuring its speed by accessing SPEED there.

In case the node is a pure v-type one, the speed associated with it may reflect the recommended speed of an electronic channel for its possible transference to other doers and can be set up, updated, or returned by SPEED too. The speed may be given by both absolute and shift values.

DOER. It returns a unique system name of the current doer, being also the name of a special node identifying the doer in the WAVE-WP model. Using DOER, it is possible to hop to the doer node directly from any other node located in it.

Assigning another name to DOER in any node (except the doer node) does not change the doer's name but rather causes transference of the current node (together with all local information and waves associated with it) to another, named doer. This is done with saving all existing links of the current node to any other nodes (in the same or any other doers). If the named doer does not exist, or transference of the node is not possible by certain reasons, the assignment will result in failure.

Assigning of a new name to DOER in the doer node itself (same as assigning to CONTENT in this node) does not also change the doer name (as it can be handled on the implementation layer only). This action rather activates the transference of all data (nodes, links, variables holding information and physical matter, interpretation infrastructures, etc.), also all passive and active waves from the current doer to another, named doer, thus adding a new load to the latter, while relieving the former. In case this complex operation is not possible, the failure will be identified.

ADDRESS. Returns a virtual address of the current node reflecting its current mapping onto EW. This address may consist of a unique (say, network) address of the current doer extended with the node's unique address within this doer. For a doer node, ADDRESS returns its network address only. Using this address, which can be assigned to other variables as their contents, the node can be accessed from any other node in the distributed system in a most efficient way.

Nodes of all kinds (including pure p-type ones), when created, are assigned mapping addresses automatically by the interpretation system, and these cannot be changed by a user from within waves. However, assigning nil to ADDRESS has a special (and severe) meaning, causing an *immediate deletion* of this node, together with all waves in it, local results, and links to other nodes. This differs drastically from the assignment of nil to CONTENT in a node (which provides graceful removal of the node, as explained before).

Concerning doer, or e-type, node, this assignment may be used as a peculiar command for an immediate shutdown of the doer. This may also be useful for simulating and testing the distributed system (on the application level) as regards its survivability and robustness after indiscriminate damages.

The mapping addresses may change over time, when nodes move between doers. So they may not represent nodes uniquely in a general case, remaining, however, the most efficient means of remote addressing of any nodes.

JOINT. This variable returns a joint position of the current node in PW, VW, and EW, as well as current system time and the node's associated speed, thus combining a response from such variables as WHERE, ADDRESS, DOER, TIME, and SPEED. By assigning to JOINT of a combined value using an aggregate constant with self-identifying components (in any order, and with some or all values to be updated), it is possible to make a simultaneous change in the current node's and wave's united space–time position. This may happen to be more efficient than accessing and updating the variables responsible for different parameters separately.

INFRA. It returns a link of the current node to the internal interpretation infrastructure supporting the current wave, making it possible to reenter the previously reached nodes by this wave in an efficient way. Using nodal variables, this link may also be recorded and subsequently used by other waves at this node either with the same color (for N-type nodal variables) or with any colors (for M-type variables).

By adding certain search parameters to the infrastructure link within an aggregate constant (like, e.g., the mentioned constants `start`, `back`, or `fringe`), INFRA (or its recorded value in a nodal variable) may allow us to access selectively (or as groups) any nodes already covered by current or other waves. This may effectively use the growing global awareness of the navigated world on the internal interpretation level.

BACK. Returns the internal infrastructure link to the previous node, that is, the one from which the current node has been reached by the current wave. This variable is functionally redundant, as similar effect may be achieved by using variable INFRA with a proper search parameter (like `-1` or `back`). But due to frequent use of the returns to direct predecessor nodes in wave algorithms, this case is supported by the special variable. Depending on the context of use, BACK may also identify and lift the network address of the previous node, via the internal infrastructure.

VALUE. Addresses the result left in the current node by the previous move, that is, by its latest operation involved. This always being a scalar value, as if the result is a vector with multiple elements, the current control point, together with the rest of wave, will be replicated (as SNR of the previous move) with each copy referring to a separate scalar. Just after hopping to any node, VALUE points at the individual content of this node (which is `nil` for p and pe types of nodes), unless it is changed by further acts in this node.

PAINT. This variable holds a value used as an identity or *color* of a wave, which accompanies waves similar to frontal variables. These colors separate identically named nodal variables produced by different waves, preventing their interference in nodes. So users with different colors may be invisible to each other in nodes by using N-type nodal variables but may communicate via M-type nodal variables, also through commonly available contents of nodes and passed links.

If different users know colors of each other, they may change (as well as restore) the value in PAINT properly, which is allowed from waves, and cooperate in full via N-type variables. Depending on the implementation, the PAINT value (or part of it) may also serve as an access permit code, possibly, of various ranks, thus granting different users different possibilities in accessing or changing certain common application or system-level resources.

RESOURCES. This may hold a vector of hardware and software resources that should (or are recommended to) be used to execute the current wave program (say, host computers, autonomous robots, special mechanical or electronic tools and gadgets, versions of operating systems, etc., as well as people).
Depending on types of resources and peculiarities of the system implementation, the resources may be physically located everywhere in the distributed world and may be shared with other waves as well as with other systems. The resources may also be appointed as a sole wave's property (even physically traveling with the waves), being loaded and activated when operations of the wave require them.
If RESOURCES holds nil, the system may be using any available resources at hand or may also be looking for them elsewhere, possibly in other systems, connecting them to the evolving waves when necessary.

USER. It allows us to have direct contact between the distributed system and the outside world. Assignment to this variable is equivalent to sending information to a terminal whatever form it might take (printer, video, audio, tactile, etc.). Reading from USER requests and waits for the online information from an outside operator, accepting from the latter arbitrary strings of characters. Depending on the implementation, USER may access either global system operators or control centers, or local terminals related to the current doers and, possibly, current nodes only.

GROUND. This variable accesses a vector of basic parameters of the current doer and the language interpreter in it, as well as physical communication channels with other doers. These parameters may relate to hardware configuration, processor and channel speed, length of the wave queue, amount of free memory, and the like.
Some parameters may be changed by assigning to GROUND, influencing system performance for a particular application on the application level. On the result of checking of different parameters, the application may come to a conclusion that another doer is needed within the situation occuring, or the current doer should be substituted for certain nodes only.

EXAMPLES OF VARIABLES

- Nodal variables: N, N345t, MM24, Mike, NONSENSE
- Frontal variables: F, FICTION, False3, FTgh567
- Environmental variables: CONTENT, RESOURCES, JOINT, VALUE

3.4 ACTS

Acts of the language are usually defined over two operands, standing on their left and right, which may generally be vectors with multiple elements. Acts can operate with multiple scalar values of the vectors simultaneously. The operands of acts may be represented explicitly, as constants or variables, or may constitute the results obtained by other acts in the same expression.

Each act (same as any wave), starting in the current node, formally produces a new SNR, with the resultant vector value distributed between these SNR nodes; it also produces a resultant control state. These results allow us to link and integrate acts with other parts of the wave program. The control states produced by acts, as already mentioned in the special constants section, can have five values: thru, out, done, fail, and abort.

Acts are divided into two classes: *flow acts* and *fusion acts*, where fusion acts may be combined with the assignment act (the latter, if standing alone, being one of the flow acts, as explained later).

$$act \longrightarrow flow\text{-}act \mid fusion\text{-}act \,[\,=\,]$$

3.4.1 Flow Acts

Flow acts permit, direct, forward, or halt data and program flow through PW, VW, and EW (or their combinations), as well as time, in the nodes they are interpreted. Flow acts may be of the following types:

$$flow\text{-}act \longrightarrow hop \mid filter \mid state\text{-}generator \mid assignment \mid code\text{-}injection$$

Let us consider the basics of their syntax and semantics.

Hops. Hops cause movement in the physical, virtual, execution, or united worlds to other (or the same) nodes. They may have the following two modifications:

$$hop \longrightarrow \# \mid \#\#$$

where act ## differs from act # in that it always causes movement to the already existing nodes, rather than to new ones, whereas act # may lead to both new and the existing nodes, with its interpretation being influenced by a context.

MOVEMENT IN PW. For movement in PW to new nodes, the right operand of act # defines the destination place coordinates, with radius (if used, see the special constants section) giving a precision of reaching the destination point (in its absence, a default precision will be used).

If the already existing PW-bound nodes are to be entered by moving in the physical world, act ## should be used instead, and the place definition should include, along with the coordinates, a radius of the search region where these nodes may be located. The left operand for movement in a pure PW should always be represented by auxiliary constant direct (possibly with additional parameters such as speed, using constant aggregation).

MOVEMENT IN VW. For movement in the virtual world, if the wave is not covered by a special rule allowing us to create networks (this will be explained later), acts # and ## behave in the same way, expressing movement to the already existing nodes, which may be done directly or through existing links between them, depending on the left operand of the hop.

The right operand supplies the name (content) or address of the node to be reached in the hop. The left operand may give the name of link(s) to be passed, in case of the surface navigation. The link name may be prefixed by a sign indicating which direction the oriented link should be passed (+ along its orientation, - opposite its orientation, and no sign—regardless of the orientation). If the link is nonoriented the sign causes no effect.

The left operand may also be a (signed or unsigned) keyword any, if all links associated with the node should be passed (with the sign reflecting their passing direction, as above). It may also be direct, if a tunnel hop to the destination node should be used, regardless of the existence of virtual links between the current and destination nodes.

If the nodes to be reached in VW may have arbitrary contents, the keyword any should be used as the right operand. If the left operand identifies a list (vector) of links to be passed from the current node, all of them will be traversed in parallel, with all nodes they lead to comprising possible destinations of the hop.

If both the left and right operands are represented by any, all nodes directly linked to the current node will be entered (some may be more than once, if there are repeating links between the nodes). And if the left operand is direct, with the right one being any, all other VW-related nodes (i.e., except the current one) will be entered throughout the whole distributed system.

Instead of any, the right operand of the hop may sometimes be represented by all, allowing all nodes to be accessed, including the current one again, and also by noback, which differs from any in that the predecessor node is excluded for the access.

MOVEMENT IN EW. In case of movement in EW between doer nodes, the right operand of # (same as ##, as doers always exist on the application level, and cannot be created by the language constructs directly) should give a unique name or address of the destination doer. The left operand may be the identifier of a physical channel between the doer nodes, or any, allowing us to use any existing channels between the current and destination doers. The destination doer in the latter case may, however, happen to be reached more than once, if there are duplicating channels between the doers.

The left operand may also be direct, accessing the needed doer regardless of the existence of physical channels between these two doers, so the doers may not be direct neighbors in the communication network, and other doers may happen to be involved in the communication process, as transit ones.

It is also possible to move from the current doer to all other doers that are connected to it, using any as the right operand. If the left operand is direct, with the right one any, all other doers throughout the whole system will be entered from the current doer. The modifications all and noback discussed before can also be used.

COMBINED HOPS. Aggregating different options within the right operand of acts # and ##, it is possible to set united integral movements in PW, VW, EW, and any of their combinations. For example, if VW nodes have presence in PW, that is, being pv (vp) or pve (vpe) nodes, it is possible to access them by their names in VW, addresses in EW, physical coordinates in PW, or any combinations of these, considerably strengthening the addressing power of separate options. The nodes can be found that, say, lie in a proper region of PW, may have listed names, and should reside in certain doers—and all this within a single hop!

Virtual links in the combined world may connect any nodes, including pure PW and EW ones. So it is possible, for example, to name virtual links at the left and move through them to PW-bound nodes defined by physical coordinates at the right, or move between EW nodes by logical, rather than physical, channels. And all the existing link addressing options can effectively combine with all the node addressing options, further increasing the integral world navigation power and expressiveness.

Time and speed may also be attached to the hops, aggregating with their left or right operands, or both (if appropriate), adding new dimensions to the world propagation. And hops may become creative, that is, forming new, persistent, nodes and links if operating within a special rule, as described later.

USING THE INFRASTRUCTURE LINK. If the node (or nodes) to be entered in a tunnel hop (by coordinates, content, or mapping address) has already been passed by the current wave or other waves with the same identity (color), and these waves are still active throughout the distributed system, using INFRA as the left operand may essentially simplify and speed up the node reentering process.

A certain parameter can be aggregated with INFRA to narrow the destination node search. This parameter may be a signed or unsigned integer reflecting the number of steps to be taken via the internal infrastructure (backward, forward, or both, the latter if unsigned), where the needed node may have been registered.

The parameter may also be a (signed or unsigned) keyword infinite, reflecting the fringe nodes of the formed hierarchical infrastructure, which can be reached by moving through the infrastructure forward, backward, or both. The infinite backward movement (with sign -) brings us to the node where the wave has started, and infinite forward movement (with sign +, using the previously recorded infrastructure link of another wave) will bring us to the SNR finally reached by that wave, and if it has not terminated yet—to all currently reached nodes by this wave at this moment in time.

Sometimes special constants start, back, and fringe can be convenient to be used with INFRA, helping to lead, correspondingly, to the node in which the whole wave started, to the predecessor node, and to the currently reached fringe nodes by this infrastructure.

Without parameters, INFRA will lead us to all nodes of the combined internal infrastructure (being or having been) automatically created by all currently active waves, which are staying or have passed the current node, these waves having the same color as the current wave.

SIMULTANEOUS HOPS. Let us consider some more details concerning simultaneous hops, where different kinds of broadcasting may occur.

If, for example, more than one link with the same name is associated with the node, and the hop uses this link name with `any` as the right operand, then all such links will be traversed simultaneously, and all nodes they lead to will be reached in parallel (some, possibly, more than once).

If `any` is used for both operands, then all links from the current node with all names will be traversed, and all nodes they lead to will be the destination nodes, entered in parallel (some nodes, again, can be replicated in the resultant SNR, if more than one link connects them with the current node).

If `direct` is used as the left operand, and the right operand is given by some name, then all nodes with this name in the network will be accessed in parallel. And if the right operand is `any`, with the left one being `direct`, then all other VW-bound nodes throughout the whole KN will be accessed in a broadcasting hop. If a node kind (or kinds) is used as the right operand, this broadcast will be narrowed to all nodes of the certain types only.

If a hop by act `##` is into a PW region given by its center's coordinates and radius, it will lead simultaneously to all PW-bound nodes lying in this region. Further destination selection may be achieved by giving the needed node kind in the right operand (more than one kind can be combined), within an aggregated value. More variants of both selective and broadcasting hops can be possible by aggregating different conditions in the right and left operands of the hop.

SPLITTING HOPS. If one or both of the left and right operands of hops `#` or `##` are vectors with more than one element (these elements may be simple or aggregate), the hop automatically splits into a set of hops with scalar operands only, at the left and right. The hops in this parallel advance, separated by commas, will correspond to all possible couplings of individual elements between the left and right vectors.

STATE, SNR, AND RESULTANT VALUE OF HOPS. Hops will produce state `thru` in all the destination nodes reached in the hop, which are comprising the hop's SNR (with resultant generalized `thru` on the hop), or `fail` on the whole hop, if no destination has been reached (with empty SNR). SNR may include (or solely consist of) the current node too. Some nodes may also be represented more than once in this SNR, as already mentioned.

The resultant value on the hop is composed of all contents of the hop's SNR nodes, as a vector. It is always `nil` if the SNR contains `p`- or `pe`-type nodes only, as these have no contents, as stated before.

If hops are in a wave that is covered by a special rule allowing new nodes and links to be created during space navigation, the semantics of hops may differ, as will be explained later.

EXAMPLES OF HOPS

- `direct # x55 y66.1`—direct hop to a new PW location in a two-dimensional region, forming a temporary node set up by absolute coordinates.

- `direct ## x77 y67 z45 r12`—direct hop to all existing PW-bound nodes located in a three-dimensional region, given by absolute coordinates of its center and a radius from this center.
- `any ## x77 y67 z45 r12`—surface hop via all existing links from the current node to all PW-bound nodes located in the region given.
- `any ## pve_x77 y67 z45 r12`—differs from the previous one in that only pve-type nodes will be entered, if any.
- `direct_dt12000 # pve_robot.99_x7 y67 z45 r12`—direct hop to all pve-type nodes residing in a particular doer (`robot.99`), if the latter is currently located in the given PW region. The movement to the needed destinations should take place within the given time shift (`12000` seconds), aggregated at the left.
- `('rival', 'friend') # ('john', 'bob')`—will be automatically converted into the following set of parallel hops with scalar operands only: `'rival' # 'john'`, `'rival' # 'bob'`, `'friend' # 'john'`, `'friend' # 'bob'`.

Filters. Filters verify certain conditions between their left and right operands, returning control states `thru` in case of a success or `fail` otherwise. Their SNR is the current node if `thru`, or empty if `fail`, with the resultant value being, correspondingly, `1` or `nil`. Filters allow us to continue or block and skip the rest of the wave program if they are standing alone as moves or are the last executed act of an expression (more details are in the expressions section).

Filters have the following options:

$$filter \longrightarrow \sim | !\sim | == | != | < | <= | > | >=$$

Their sense and main use are as follows.

Belongs ($\sim$) checks whether the vector at the left as a set (i.e., all its scalar values) belongs to the vector at the right, as to another set. This act fails if at least one value from the left does not belong to the vector at the right. The fact that both vectors may have repeating elements in general (as explained before) does not influence the result of this act.

Does not belong (`!`$\sim$) checks whether the vector at the left (its every scalar value) does not belong to the vector at the right. The act fails if at least one such value belongs to the right vector.

Equal (`==`) succeeds only if the two vectors are exactly the same, including the number of elements, their order, and values of peers.

Not equal (`!=`) succeeds if there is any difference between the vectors at the left and the right, that is, in the number of elements, their order, and values of peers.

Less (`<`), **less-or-equal** (`<=`), **greater** (`>`), and **greater-or-equal** (`>=`) succeed only if every scalar of the left vector is in the named relation with the peer scalar of the right vector, otherwise general failure occurs. The shorter vector is automatically extended to the size of the longer one by zeros or empty strings, depending on the form of the extra elements of the longer operand. If peer scalars cannot be treated as numbers (at least one of them), they are compared as strings (in the way they are written), using lexicographical order for the comparison.

EXAMPLES OF FILTERS

- (1, 1, 2, 5) ~ (5, 4, 1, 7, 2)—will result with thru.
- (6, 33, 47, 14.6, 25) > (3, 25, 13)—the right operand will be extended by zeros as (3, 25, 13, 0, 0), and the act will produce state thru.
- (44, 22, 35) != (44.00, 22, 35.0)—will result with state fail, as each element in the left vector has the same numeric value as the peer element in the right vector.

State Generator. The state generator uses ! as its act symbol:

$$state\text{-}generator \longrightarrow \ !$$

The right operand of this act sets explicit control states that, listed in priority, may be of the four values: abort, thru, out, done, or fail, as already mentioned.

State abort immediately terminates the whole wave, regardless of the left operand. This removes all processes and temporary data produced by the current program, which are associated not only with the current node where this state has emerged but also with all other nodes the wave navigated by this time, by all its branches. The scope of influence of this emergency state may be limited, if needed, to the part of the wave program covered by a special rule, this part containing the abort-generating act (see more in rules).

State thru identifies full success, treating the current node as the act's SNR and the value of the left operand as the resultant value of the act. The rest of the program, if any, will work from the current node. State out has the same effect as thru if it appears inside a sequence of moves (advances) but is supplied with additional semantics for the management of cyclic parts of the program, as explained later.

State done signals successful premature termination of the wave. With it, act ! gives the same resultant value as with thru (by the left operand) and SNR as the current node. It also makes done and the current node, correspondingly, the final state and final SNR on the whole wave program in this node, thus skipping the rest of wave, also prohibiting further development of higher level waves from this node.

State fail indicates the wave's failure in the current node, with act ! resulting in an empty SNR and nil value, regardless of the value of the left operand.

As can be seen, only states thru (and its modification out if used for this) permit further forward wave processes from the current node, whereas thru, out, and done allow us to have and collect the resultant value of the wave (see the expressions section). States thru, out, and done in WAVE-WP may be considered as three variations of the more traditional *true*, whereas fail corresponds to traditional *false*.

States thru and fail are also generated automatically during execution of other acts, as an embedded indication of their success or failure. The other three, out, abort, and done, can be produced only explicitly, by the act !. All these five states can be effectively used on higher language levels for making automatic decisions within the spatial control managed by rules (see later).

EXAMPLES OF STATE GENERATION

- `beta`! thru—signals success and allows the rest of the wave, if any, to continue from the current node. The resultant value of this act will be the one shown at the left of !. This wave will, however, be equivalent to just representing the left operand as an independent move: `beta`, as state thru for this, will be generated automatically on the internal level.
- `beta`! done—represents a planned successful termination. The rest of the wave, if any, will be skipped. The resultant value will be `beta`, which can be used in higher level expressions. As already mentioned, state done can be generated only explicitly, by act !.

Assignment. Assignment is expressed as usual:

$$assignment \longrightarrow =$$

It records the result obtained at the right (which may have multiple elements, as a vector) to the left operand. The main purpose of the left operand, which may itself represent any wave, is to establish a link (or links, as there may be more than one) to certain node contents or to the already existing or to be created variables (which may be local, in the current node, or remote, in any other nodes).

These variables may be indexed or contented, as explained later. The contents of all reached nodes or the variables in them will be updated simultaneously by the same assignment act, using the same value of the right operand.

The SNR of the assignment is always the current node, control state is always thru, and the resultant value is always nil. The latter is because the language semantics considers the assignment of the result (to variables or nodes) as hiding this result, formally relieving the current node from the result's value.

EXAMPLES OF ASSIGNMENTS

- N = 5—a usual assignment of constant 5 to nodal variable N, while staying in the current node, with the resultant state thru and resultant value nil.
- (`link1` # `node1`; N) = (`link2` # `node2`; N1)—assignment of the content of nodal variable N1 to nodal variable N, where both variables are in other nodes: node2 and node1, which are accessible from the current node, respectively, by links named link2 and link1.
- (`link1` # `node1`) = (`link2` # `node2`)—instead of variables, the contents of the nodes will be engaged, with substituting node1 by node2 as the content of the first node, so now two nodes with the same content (name) node2 will exist in the virtual network.

- (`'link1' # any`) = (`'link2' # any`)—differs from the previous case in that node names (same as contents) reached by the named links may be arbitrary, and the first node will change its original content to the content of the second node, whatever these contents might be.

Code Injection. This operation has two options, with act symbols being as follows:

$$code\text{-}injection \longrightarrow \; \hat{} \; | \; ?$$

All string values provided by the right operand of act ^ (which may generally be a vector) are treated as wave programs. They will be injected into the current wave as a new part of its body and executed in parallel. The left operand supplies the names of frontal variables of the current wave that may potentially be used in the injected programs.

The SNR of act ^ is an amalgamation of final SNRs of all injected programs, and the resultant value of the act comprises all values obtained in this SNR. The control state of the act will be represented as the generalization of all final states in its SNR.

The act ? differs from the previous act ^ in that it activates external procedures and systems, possibly, in other languages (supposed to be recognized automatically by the interpretation system). The right operand of this act may generally be a vector of other systems calls, and the left operand may provide actual parameters for invocation of these external procedures.

EXAMPLES OF CODE INJECTION

- (`F, F2, F5`)^(`N1, FF33, 'link25' # any`)—new waves (being the string contents of nodal variable `N1`, frontal variable `FF33`, and the node reached from the current node by the link named `link25`) will be injected into the current wave as independent moves and executed concurrently. These new waves may use for their work the three frontal variables named in the left operand of ^ to carry transitional data through the space they will navigate.
- `'mobile agents' ? ('www.google.com', 'www.yahoo.com')`—will activate independently the two Internet search engines, which will collect data on the `mobile agents` topic, and will associate the obtained (multiple) results with the current node, as a resultant vector.

3.4.2 Fusion Acts

. Fusion acts provide different sorts of data processing, which may produce new data from other data. These acts may be of the following types:

$$fusion\text{-}act \longrightarrow arithmetic \; | \; vector\text{-}special \; | \; composition$$

The left and right operands of the fusion acts may, in general, have multiple values, as vectors, producing vector results too. Fusion acts usually terminate with control state `thru` (with a few exceptions explained later), and the current node as their SNR with the resultant value in it.

If the result of acts has multiple scalar values, each of them will be associated with the separate copy of control in the current node, these copies identifying altogether the act's SNR. This enables the following move, if it exists, to replicate and process all elements of the obtained vector *independently and in parallel.*

As already mentioned, fusion acts may be combined with the assignment act, with the left operand now used for both supplying data for the act (along with the right operand) and as a storing destination (or destinations, which may be multiple) for the result obtained. This may save a good deal of program code and save time (as the left operand may be arbitrary complex wave spreading in distributed space, and it will be executed only once now).

Arithmetic Acts. These have the following representations:

$$arithmetic \longrightarrow + \,|\, - \,|\, * \,|\, / \,|\, **$$

Arithmetic acts operate on their left and right vectors, with operations performed *between peer elements only*, with the resultant vector having the number of elements corresponding to the longer vector. Missing elements of the shorter vector are extended by zeroes. This also covers the case where one operand is `nil` (or missing): It will be represented by all zeroes.

If any vector element is a string in single quotes, it is to be treated as a number. If this is impossible, the result of the whole act (regardless of the other scalars) will be `nil`, control state `false`, and SNR empty.

The general semantics of the arithmetic acts is as follows:

Addition (+) adds two vectors.
Subtraction (−) subtracts the right vector from the left one.
Multiplication (*) multiplies two vectors.
Division (/) divides the left vector by the right one. If the right vector originally has (or is extended by) zero elements, the whole act will terminate with state `fail`, empty SNR, and `nil` value.
Degree (**) puts the left operand into a degree given by the right operand.

Examples of Arithmetic Acts

- (1.2, 3.5, 2, 5.2) + (3.1, 6.0, 7.7)—after extending the right vector by a zero scalar, this will result in (4.3, 9.5, 9.7, 5.2), with each obtained scalar formally associated with a separate copy of control in the current node, altogether forming the resultant SNR.
- (6, 8) − (1, 2, 3, 4)—will result in (5, 6, −3, −4) after adding two zero elements to the first vector, each scalar with a separate control copy of the current node.

- (5, 4, 3) * (3, 4, 5, 7, 9)—will give (15, 16, 15, 0, 0), each scalar associated with a separate control copy of the current node.
- (8, 5, 1, 3)/(4, 6, 1)—will result with nil value, state fail, and empty SNR, as the extended right operand will contain a zero by which scalar 3 of the left operand will fail to be divided.

Special Vector Acts. These acts have the following representations:

$$vector\text{-}special \longrightarrow \& \mid : \mid ::$$

They either assemble new vectors from other vectors or provide different forms of an access to vector elements. Resultant state of these acts is always thru, regardless of the resultant value, which may be a vector.

Append (&) attaches a vector at the right to the end of a vector at the left, forming a united vector, in which elements may be of mixed types.

Indexing (:) points at elements in the left vector by indices given in the right vector, with selected scalars from the left operand forming the resultant vector. If the pointed scalars at the left are used in a subsequent assignment act as its own left operand, each of them will be substituted by the same value from the right hand of the assignment, which itself may be a vector. So the vector at the left of the indexing act, after the following assignment, may increase or decrease in size (the increase will take place if more than one scalar will substitute single scalars, and decrease—when the assigned value is nil).

Contenting (::) points at elements in the left vector by their contents given in the right vector, with the indices of the pointed scalars forming the resultant vector. In case of a subsequent assignment in the expression, similar to the indexing act, the pointed elements will be substituted by all values from the right hand of the assignment.

EXAMPLES OF SPECIAL VECTOR ACTS

- (a, b, c, d) & (1, 2, 3, 4)—will result in (a, b, c, d, 1, 2, 3, 4).
- (a, b, c, d) : (1, 1, 6, 4, 2, 2)—will result in (a, a, d, b, b), with the nonexisting index 6 returning nil value skipped in the result.
- (a, b, c, d) :: (b, b, b, c, a)—will give (2, 2, 2, 3, 1).
- (a, b, c, d) : 8—will result with nil.
- N = (a, b, c, d); N : (1, 3)=(new1, new2)—will result in N: (new1, new2, b, new1, new2, d).
- N = (a, b, c, d); N :: (a, c)=(new1, new2)—will give the same result as in the previous case.

Composition–Decomposition. The composition–decomposition (shortened to *composition*) acts have the following representations (one of them is shown in bold to distinguish it from the delimiter used in the syntax description):

$$composition \longrightarrow \mathbf{|} \mid \mathbf{\%} \mid \mathbf{\&\&} \mid \mathbf{\_}$$

These acts allow us to decompose different kinds of objects into proper elements or compose new objects from their parts. The resultant state for them is always thru, regardless of the resultant value.

Splitting (|) partitions an object at the left (which may be of any type described in the constants section) into a vector of elements using a delimiter given at the right. If the left operand is itself a vector with more than one element, each such element will be split by the same delimiter, and the obtained vectors will be appended to each other in the original order, to form a single united vector.

If the right operand is a vector with more than one element, these elements will be used as delimiters in a round robin fashion, with each element used in full, as before, and with the final result represented as a single united vector too, with intermediate vectors appended to each other in the proper sequence.

Merging (%) glues together elements of the vector at the left into a united object, with a delimiter between the elements given by the right operand. If the latter is a vector with more than one element, they will be used one after the other in a round robin fashion. If element '!' is among the delimiters, it terminates the current object and starts a new one, with the obtained objects forming the resultant vector in the original order.

Concatenation (&&) attaches the value of the right operand to the value of the left one without any delimiter, forming the new (string) object. If operands are vectors, this operation is performed on their peer scalars only, with the final result being a vector of the assembled objects (the shorter vector of the two is extended by empty string elements).

Aggregation (_) forms a single, aggregate, value by linking the left and right operands with the delimiter _ (same as the act symbol itself!). If the operands are vectors, this operation is performed between their peer scalars, with the result being a vector of aggregate values. If the right vector is shorter, its elements are used in a round robin fashion to match all elements of the left operand. If the right operand is longer, its exceeding elements will be skipped.

EXAMPLES OF COMPOSITION–DECOMPOSITION

- 'WAVE is the language for network processing'|' '—will result in the vector (WAVE, is, the, language, for, network, processing).
- 'WAVE is the language for network processing'|('is the', 'for network')—will give (WAVE, language, processing).
- (WAVE, is, the, language, for, network, processing) % (' ', '', '', !)—will result in the vector consisting of two scalars ('WAVE is the language', 'for network processing').
- (a, b, c, d) && (p, q, r)—will result in (ap, bq, cr, d).

- (`peter`, `paul`, `john`, `nick`)_(robot.1, robot.2)—will result in the vector (`peter`_robot.1, `paul`_robot.2, `john`_robot.1, `nick`_robot.2).

3.5 RULES

3.5.1 Rules in General

Rules establish special constraints and contexts over waves and their distributed development in space. They may also influence the resultant SNR of the wave they embrace, which may differ from the SNR of this wave when the latter develops free (being, however, based on this SNR).

Most rules provide spatial synchronization of the embraced wave, allowing the rest of the program to develop from the rule's resultant SNR only after full completion of the ruled wave. This synchronization is usually logical rather than physical, as waves may develop in fully distributed and asynchronous manner. The top control over the development of rules in distributed space is always associated with the node where the rule has started.

Rules may belong to the following two classes:

$$rule \longrightarrow forward\text{-}rule \mid echo\text{-}rule$$

Forward rules, to which most of the above relates, first of all coordinate world navigation and spreading of activities in distributed spaces. The resultant value of the forward rules comprises the final values obtained (or reached) in their SNR.

Echo rules return results, obtained by the wave they embrace, to the current node, that is, the node in which they started, after a certain processing or conversion *during*, rather than *after*, their distributed collection and return. SNR on these rules is always represented by the current node (not by SNR or a derivative from it of the wave they embrace, like for forward rules). The resultant control state is always thru for them (if abort did not happen in the ruled wave).

If the returned result by an echo rule is a vector with more than one element, each scalar will be formally associated with a separate copy of the current node, as usual, and these copies will comprise the resultant SNR on the rule rather than the single current node.

Let us learn more about the semantics of different rules from these two classes and how they can be used for structuring of complex waves navigating distributed worlds.

3.5.2 State Generalization Procedure

A number of forward rules makes different decisions on the basis of a generalized termination state of the wave they control (or parts of this wave). These rules produce their individual termination state, which may differ from the termination state, of the wave they rule, being, however, a derivative of this state. The same may be said about their resultant SNR, which is generally a derivative of the ruled wave's SNR.

Generalized termination state on any wave without rules is determined as the *maximum of local termination states* produced by its different branches in final nodes reached, with `thru` being the strongest and `fail` the weakest states. If rules are present and form a hierarchy, possibly, a recursive one, the state generalization becomes hierarchical (and recursive) too, where the individual termination state on a rule is taken into account on higher levels rather than the state on the wave it embraces.

The state `abort`, immediately terminating all distributed processes, is beyond this state generalization scheme but may be controlled by a special rule discussed later.

3.6 FORWARD RULES

Forward rules may be of the following types:

forward-rule $\longrightarrow$ *branching* | *repetition* | *synchronization* | *resource-protection* | *network-creation* | *autonomy*

3.6.1 Branching Rules

Branching rules *split* the wave into separate branches and coordinate their parallel or sequential development in space and time. The rules return their personal termination state (`thru`, `done`, or `fail`), which is a derivative from the termination states of different branches. Each branching rule also produces a peculiar personal SNR based on SNRs of the branches formed. And the resultant value on branching rules comprises the values obtained in all its SNR nodes.

The following branching rules are present in the language:

branching $\longrightarrow$ `sequence` | `or` | `orparallel` | `and` | `andparallel` | `random`

Splitting into Branches. The semantics of splitting into branches is hierarchical and recursive, and the splitting starts on the top syntactic level, proceeding down until successful. The following steps are taking place in the splitting process.

1. If the *head* of wave (i.e., first advance separated from the rest of wave by a semicolon) consists of more than one move on top level (i.e., not embraced by rules or just parentheses, with moves separated by commas), the branches are formed starting with these moves, *before* execution of the head. The *rest of wave*, if any, will be replicated and attached to each of these moves, completing the branch.

2. If the head is a single hop or code injection act, the following options will be examined, with branches formed *during* execution of the head, if this is possible.

- If the hop has vector operands with more than one element, thus splitting into simple hops with scalar operands (as explained before), these simple hops, with the rest of wave following each of them, will be considered as branches.
- If the hop is simple but broadcasts to other nodes (see the hop act section), each individual hop emerging within this broadcast will start a *dynamic branch*, with the rest of the wave following these individual hops.
- If a vector of procedures is injected by the acts ^ or ?, each such procedure, followed by the replicated rest of the wave, will form a separate branch.
- If only a single procedure is injected, the injected code, together with the rest of the wave attached to it, will be considered as a completely new wave embraced by the current branching rule, and the splitting will start from the beginning, that is, from step 1.

3. If the head of the wave, due to recursion, is itself a ruled (or at least parenthesized) wave providing after its execution an SNR with more than one node, the branches will start in each such node *after* execution of the head. These branches, starting in *different nodes*, will consist of the rest of the wave only.

4. The embraced wave will be declared as a single branch only, as all the above steps failed.

Branching Rules Semantics

sequence Executes the formed branches sequentially, one after the other, in the order written or formed in the splitting process. Each new branch is launched only after full termination of the previous branch, regardless of the success or failure of the latter. A final SNR on the rule is composed of SNRs received on all branches, and the personal termination state of the rule is the maximum of the termination states of all branches.

or Executes the branches sequentially too but terminates with the first successful branch—that is, the one that returns generalized state thru or done. SNR on this branch and its state form, correspondingly, the SNR and state of the rule. If no branch replies with a success, the whole construct will result with state fail, empty SNR, and resultant value nil.

orparallel Contrary to the sequential or, activates all branches in parallel, terminating immediately as a whole if a branch replies with termination state thru or done, taking SNR, state, and value produced on this branch as the results of the rule. All other activated branches will be aborted without waiting for completion. If no branch replies with a success, the whole construct will result with state fail, empty SNR, and nil value.

and Executes branches sequentially, terminating with fail (also empty SNR and nil value) of the whole construct if a current branch results with fail. Only if all branches terminate with success (i.e., thru or done), the result of the rule will be success too (with termination state thru if at least one branch replied with thru, otherwise done). The resultant SNR on the rule will be the union of SNRs on all branches (similar to sequence, where, however, branches may result with fail too).

andparallel Differs from the previous rule by activating all branches in parallel, with the resultant state and SNR formed similarly. If any branch results with generalized `fail`, the whole construct will immediately terminate with `fail` too, empty SNR, and `nil` value, aborting all branches that may still be in progress.

random Chooses randomly only one branch among the branches formed by the splitting process and executes it, with the rule's termination state, SNR, and resultant value inheriting those of the branch chosen. This rule differs from the rules `or` and `orparallel` in that the resultant branch for `random` is chosen *before*, not *after*, its execution, and this branch may not necessarily lead to a success.

EXAMPLES OF BRANCHING RULES. In the following examples, `w1` and `w2` may be arbitrary waves.

- `sequence('link1' # any, 'link2' # any, N22 = 5); w1`—executes sequentially the three moves (each starting from the same, current, node). The first two moves are hops to other nodes via links named `'link1'` and `'link2'`, and the third one is an assignment in the current node. The resultant SNR will comprise the nodes reached by these links, as well as the current node, from all of which the remainder of the program, `w1`, will be developing independently and in parallel. The application of `w1` will be logically synchronized, waiting until the whole construct `sequence` terminates completely.
- `sequence('link1' # any, 'link2' # any, N22 = 5; w1)`—will be automatically converted into `sequence(('link1' # any; w1), ('link2' # any; w1), (N22 = 5; w1))`, creating three branches by splitting the head and replicating and attaching the tail `w1`, which will now become the property of each branch. The remainder of the wave, if any, will be applied in parallel to all nodes belonging to the united SNR on all these branches, and global synchronization will take place *after* completion of all invocations of `w1`, *not before* as in the previous example.
- `orparallel('link' # any; w1)`—will create branches dynamically, if the hop provides broadcasting to more than one node from the current node by all links named `'link'`. Each such elementary hop, followed by the same tail `w1`, will be executed as an independent branch, in parallel with other branches. The control state, SNR, and resultant value will be defined by the first branch that replied with a success (i.e., `thru` or `done`). If no branch replies with a success, the resultant state will be `fail`, SNR empty, and resultant value `nil`.
- `and(or(w1); w2)`—the head of the ruled wave in this example is itself a ruled wave. All branches will be formed from the same tail `w2` and will start in all nodes of SNR provided by the head `or(w1)`, after its execution, so the number of branches may not be known in advance and will correspond to the number of nodes in this SNR. The branches will be executed sequentially. An overall failure will occur if a branch fails. The resultant value on the rule will comprise the values obtained on all branches.

3.6.2 Repetition

This rule has the following representation:

$$repetition \longrightarrow \texttt{repeat}$$

It first allows the embraced wave to develop in a regular way from the current node, and if it results with control state `thru`, and therefore nonempty SNR, the `repeat` construct will be applied again independently, asynchronously, and in parallel from all the SNR nodes, each such invocation producing new SNR, to which `repeat` will be applied again, and so on.

If the ruled wave starting from a current node results with `fail`, the `repeat` will stop in this node, and the current node will be included into the final SNR on the whole `repeat` construct. The value obtained in this node prior to the failed application of `repeat` will be included into the final (multiple) value on the whole `repeat`.

If the ruled wave returns generalized state `done`, the current invocation of `repeat` will stop, and the nodes in which `done` occurred, along with values in them, will be included, respectively, into the SNR and value on the whole rule. But the rest of wave, if any, will be prohibited to develop from the nodes with occurred state `done`, as usual.

Some more details on this rule are as follows. If the ruled wave returns the generalized state `out`, the `repeat` construct terminates, and the rest of the wave will continue from the reached nodes with this state, rather than from the current node (i.e., the nodes with `out` will be included into the final SNR on the whole `repeat` construct).

In the mixed case, if the ruled wave returns generalized `thru` as the most superior one on all possible states, but some branches terminate with `done` or `out`, the `repeat` will continue from the nodes with `thru`, the whole wave will terminate in nodes with `done`, and the rest of the wave will develop from the nodes with `out` (which are included into the final SNR on `repeat`).

EXAMPLES OF REPETITION

- `repeat(- 'partof' # any)`—starting from some node, repeatedly moves up the hierarchy via the oriented links named `'part of'`, opposite their orientation, until the top of hierarchy is reached, in which the remainder of the program, if any, will be applied. It is assumed for this example that links in the hierarchy are oriented top-down, so the resultant SNR on `repeat` will correspond to the top node, with the node's content as the resultant value.
- `repeat(+ 'part of' # any)`—starting from some node, repeats moving down the same hierarchy, along the orientation of its links, navigating the hierarchy in parallel (the nodes may have multiple successor nodes). The SNR on this program will be the fringe nodes reached, in which the navigation will stop, with the resultant value comprising all contents of these nodes.

- `repeat(N+=1; N<=100; w1)`—repeats wave `w1` (assumed always leaving control in the same, current, node) a hundred times, after which the remainder of the program, if any, will continue from the current node (the latter being the final SNR on `repeat`, with final value in the node as the value obtained by last invocation of `w1`).

3.6.3 Synchronization

Synchronization rule has the following representation:

$$synchronization \longrightarrow \texttt{wait}$$

This rule, always having SNR, state, and value of the wave it embraces, allows the rest of the program to be applied in its SNR only after the wave's complete termination, similar to other rules, which have different additional semantics. The rule `wait` has a peculiar payload too, managing the reaction on state `abort`, as follows.

If `abort` occurs within the wave controlled with `wait`, the whole wave, which may be evolving in parallel and distributed mode, will be immediately (actually as soon as possible in the distributed environment) terminated, with all nodal variables and internal infrastructures produced by it removed. The rule itself, however, will result with state `fail`, not `abort` (along with empty SNR and `nil` value), thus containing the emergency abortion within the wave it controls. For an outside control, this will be treated as a usual failure. In case `abort` occurs outside the scope of `wait`, the whole distributed program will be aborted, as mentioned before.

EXAMPLE OF SYNCHRONIZATION

- `wait (w1, w2, w3); w4`—synchronizes termination of the three waves `w1`, `w2`, and `w3`, after which `w4` will start simultaneously in all nodes reached in space by the three waves. If in `w1`, `w2`, or `w3` state `abort` occurs, all these waves will terminate immediately together with all data and internal infrastructures created by them, with `wait` producing state `fail`, empty SNR, and `nil` value. Wave `w4` will be skipped.

3.6.4 Protecting Common Resources

This rule has the following representation:

$$resource\text{-}protection \longrightarrow \texttt{grasp}$$

Applied in a node, it establishes an absolute control over all the node's resources (like node content and nodal variables), which will be available now from the ruled wave only. This blocks activity of other waves in this node until after full completion of the current wave. The ruled wave can be arbitrary and can cover other nodes too.

If applied in a doer node, `grasp` blocks other waves not only in this personal current doer node but also in all other nodes located in the current doer, until the embraced wave terminates. The control state, SNR, and final value produced on `grasp` are the same as for the wave it covers.

If the wave under `grasp`, starting from the current node, can cover other nodes, and if in these nodes another `grasp` is simultaneously applied with its own wave capable of covering the current node, a system deadlock can occur, as neither rule may happen to be able to complete. So extending the scope of application of this rule to other nodes throughout the network should be done with care.

EXAMPLE OF RESOURCE PROTECTION

- `grasp(F < N; N = F)`—will allow us to execute the filter and assignment acts only after receiving permission to have a sole access to all resources associated with the current node. Having started, `grasp` blocks other activities in the current node until after all operations under this rule terminate.

3.6.5 Network Creation

This rule has the following representation:

$$\textit{network-creation} \longrightarrow \texttt{create}$$

The rule supplies the wave with a possibility of creating new VW-related nodes (i.e., of the types v, pv or vp, and pve or vpe) and virtual links between any types of nodes, rather than traversing the existing nodes and links if the latter exist. Rule `create` influences only hops in the ruled wave, whereas other acts and rules operate as usual. Some details are as follows.

If the link operand in the hop under `create` is given by the link's name, a new link will be created (oriented if prefixed by a sign, otherwise nonoriented), with the following node details.

- the destination node is given by the its name in VW or coordinates in PW, the new link–node pair will be created if hop # is used.
- If the node is represented by its address (indicating that it already exists, as addresses are assigned automatically), only a new link to the existing node will be created.
- If hop ## is used with any node definition at the right (i.e., by name, address, coordinates, kind, any, or their combinations) only a link to the existing node (or nodes) will be created.

If `direct` for the link operand is used, an isolated VW-related node will be created by act #, if the node's name is given by the right operand.

As explained before, for movement in pure PW by coordinates, act # always forms new, temporary nodes without the need to use `create`, as PW is always considered existing independently. Using `create` as a context, however, makes it possible to form new PW nodes together with virtual links leading to them from the current node, if act # is used, or establish only new links to the existing PW nodes by act ##.

In case a hop to existing nodes is given by a construct leading to more than one node (using, say, node kind, `any`, name of a doer the nodes can stay in, or a region in PW where they can be located, as well as combination thereof), links with the same name will be created from the current node to all such nodes.

Other variations of hops, without explicit naming of both nodes and links, will remain unaffected by rule `create` and will operate as usual.

Using this rule allows us to create the whole networks by integral patterns "in a single breath," rather than by separate pieces, as usual. The same patterns in the language syntax can be used for both navigation of the existing networks and creation of new networks (if just enclosed by `create` in the latter case).

EXAMPLES OF NETWORK CREATION

- `create(direct # 'a'; + 'p' # 'b')`—creates VW node a, and then node b, into which oriented link p will lead from a. The rest of wave, if any, will be applied in node b.
- `create(-'q' ## 'a')`—creates only oriented link q to the already existing VW-bound node a from the current node, this link will be oriented from node a into the current node. The rest of wave, if any, will be applied in node a.
- `create(+'control' ## x44 y55 r3)`—creates oriented links named `control` leading to all existing PW-bound nodes located in the region defined by coordinates of its center and radius. The remainder of wave, if any, will be applied in all the nodes reached.

3.6.6 Autonomy Rules

These rules relax the originally strong hierarchical and chainlike control of the evolution of waves in space, being of the following three types:

$$autonomy \longrightarrow \texttt{release} \,|\, \texttt{free} \,|\, \texttt{quit}$$

release. Grants full independence to the wave it embraces, letting the latter develop from the current node without subordination to the main program that launched it. The main program immediately considers the released wave as terminated successfully (with state `thru`), without waiting for its actual termination. And `release` results for the main program with SNR that comprises the current node only, with the resultant value that was in the current node before the rule's application. The rest of wave will be developing from the current node as if the ruled wave did not exist.

The released wave will behave as a totally independent program, like having been injected from the very beginning in the current node.

free. Removes hierarchical, chained in distributed space, command and control from the ruled wave on the internal level, freeing each new advance from its control dependence from the previous advance, and not reporting its status back. It is equivalent, in some sense, to a repeated invocation of the previous rule `release` before each advance.

The chained hierarchical control will, however, be switched on automatically if any other rule is met in the wave in nodes. The renewed spatial control will hold until this rule terminates, after which the free propagation will continue. The fate of N-type colored nodal variables produced under `free`, after leaving the nodes, is undefined; they may be removed, as litter, by the interpretation system after an expiration of a threshold time.

The resultant state, SNR, and value on this rule, in relation to the main program that launched it, correspond to these of the previous rule `release`. The main program will consider the wave under `free` as terminated immediately too.

quit. First allows the embraced wave to develop and complete in a regular way, as an integral part of the main program. But upon the rule's termination, it immediately removes all distributed temporary information produced by this wave, including colored nodal variables and internal interpretation infrastructures.

This means it prematurely launches distributed cleanup procedure usually involved only after full termination of the whole wave program. For the main program, `quit` always results with an empty SNR, `nil` value, and state `done`, if the embraced wave terminated with `thru` or `done`, otherwise with `fail`. So the rest of the wave, if any, will always be blocked and skipped.

EXAMPLES OF AUTONOMY RULES

- `release(w1); w2`—allows `w1` to develop as an independent program and `w2` immediately starts from the current node.
- `free(w1; or(w2, w3); w4)`—will make `w1` develop free, then sequential and synchronizing `or`-type control will cover `w2` and `w3`, after which `w4` will develop free again.
- `quit(repeat(any # any; grasp(N==nil; N=1; USER=CONTENT)))` —removes all results of the embraced wave immediately after its termination, not waiting for the termination of the whole wave program. (The repetitive wave under `quit` navigates the system in a breadth-first parallel mode using nodal variable `N` to protect nodes from entering them more than once, printing the names of nodes.)

3.7 ECHO RULES

Echo rules, from the implementation point of view, allow the resultant values to be obtained and brought to the current node in a most efficient way, by a parallel and distributed echo-merging process, possibly, using resources in other computers, which have been navigated in the forward process. Examples of some possible echo rules are shown below, whereas their full collection will depend on the expected application areas and system implementation.

$$echo\text{-}rule \longrightarrow \texttt{state} \mid \texttt{rake} \mid \texttt{min} \mid \texttt{max} \mid \texttt{sort} \mid \texttt{sum} \mid \texttt{product} \mid \texttt{count} \mid \texttt{none}$$

The semantics of these rules is in brief as follows.

state Returns the resultant generalized termination state of the embraced wave as a resultant *value*, using numbers 2, 1, and 0 representing, correspondingly, thru, done, and fail. Usually these states are generalized and returned on the implementation level implicitly, as *states*, not values, and influence the work of branching rules. Rule state makes these resultant states visible, and this may simplify making complex system decisions, also assist in composing new rules by using the existing rules. State abort (equivalent to 3) cannot be returned by state, as it terminates and removes the whole program, including all rules in it.

rake Collects the results obtained on the wave (which may be multiple and remote) in an *unordered* manner. The collection, for example, may follow a natural order of their appearance, as well as movement (with unpredictable delays) up the distributed hierarchy. This may differ from the written order in a wave or the order in which dynamic forward processes were evolving, which is strictly obeyed when the results are returned from waves without rake.

min, max Find, respectively, minimum and maximum values among all the values reached or obtained by the wave, which may be remote.

sort Collects the results obtained by the wave in the ascending or descending (as chosen) order of their values.

sum, product Respectively, sums up or mutually multiplies all the results obtained by the wave.

count Gives the number of elements in the result.

none Returns no results, that is, having nil as a value, just letting the embraced wave develop from the current node and terminate as usual, without returning the results reached or obtained, and then apply the rest of the wave from the current node, not from the SNR obtained on the wave. Rule none is especially convenient for simplification of structuring of parallel and distributed wave programs. Instead of this rule, just square brackets embracing the corresponding wave may be used.

EXAMPLES OF ECHO RULES

- min(state(w1), state(w2), state(w3))==1; w4—will activate wave w4 from the current node only if the minimum termination state on waves w1, w2, and w3 will equal 1 (i.e., done). This will be inconvenient to express using other rules of the language.
- N44 = sort(3, 56, (w3; w4; Nn), repeat(+ 'link' #))—assuming that w3 and w4 are arbitrary waves, and a single Nn variable reached in a

remote node has vector content (5, 15, 22), and finally reached nodes after repeatedly moving from the current node along oriented links named `link` have contents 88, 29, and 105, the resultant value in N44 in the current node will be (105, 88, 56, 29, 22, 15, 5, 3). This scattered data collection and sorting will be performed in parallel, using processing facilities associated with different nodes.

3.8 EXPRESSIONS

Acts, operating on their left and right operands, may be combined in expressions, where the results obtained by one act may be used by other act, as their operands. Any sequence of acts is allowed in an expression, and acts are generally executed in the written order, each operating on its left and right operands. This order may, however, change if both fusion and flow acts are in the same expression, as *fusion acts are superior to flow acts and must be executed first.* The general left-to-right execution order can be changed if parentheses are used for structuring of the expressions: The waves in them will have a priority and should be processed first.

Both left and right operands of acts in an expression may be constants, variables, or results obtained by other acts (or their groups, in parentheses). Taking also into account that parenthesized parts may be arbitrary waves, with their SNR located anywhere in the distributed world, the operands of acts may generally be vectors with multiple and remote values.

Conceptually, it is assumed that all acts of the expression are always performed in the current node, that is, where the expression has started (except, possibly, acts of parenthesized parts of the expression, which may belong to arbitrary waves and therefore may potentially be executed in other nodes). And values of both the left and right operands of acts are assumed to be delivered to the current node before the act's execution. Also assumed is that the result of acts is left in the current node. This, however, may not hold if the executed act is a hop or code injection, both potentially leading to other nodes.

Acts in an expression are executed regardless of the control states of the previous acts in this expression (except `abort`), as what is used from the previous acts is only the value they produce (which may be `nil` too, along with state `fail`). The final SNR, resultant value in this SNR, and resultant control state of an expression are those of its *last executed act* (which may differ from the *last written act*, as the execution order may depend on the existing subordination between flow and fusion acts and also presence of parentheses).

EXAMPLES OF EXPRESSIONS

- 56.7, Next, M44 = 56 + 234.1 - F66—elementary expressions.
- w1 # ('peter', 'paul'); w2—where # is a hop act, w1 results in link information (which may be remote), and peter and paul being the names of

destination nodes. Wave w2 will develop from the reached nodes peter and paul independently and in parallel.
- (create('link1' # 'node1'; N1), ('link2' # any; N2:4))= ('link3' # any) * ('link4' # any; 'link5' # any; N3)—this is an example of the expression with remote variable N1 (in the newly created node named node1, reached by new link named link1) and indexed remote variable N2 (in the node reached by link named link2). These variables will be updated by the result obtained at the right hand of the assignment. This result is a multiplication of the content of a node reached by link named link3 on the value of variable N3 in the node reached via link named link4 followed by link5. Nodes accessed by link3, and separately by the sequence of link4 and link5, may be multiple, so the final result, as a potential vector, will be assigned to variable N1 and will substitute the fourth element of the vector in variable N2.

3.9 WORKING WITH PHYSICAL MATTER

We have considered the work of WAVE-WP acts with information only. Some of these acts can formally be applied to physical matter or physical objects and their storage too, while many more specific acts or special external functions may be needed to operate in the physical world directly.

When working with physical matter or physical objects, it should be taken into account that it may not always be possible to copy or replicate them like information. Some operations on them can preserve and leave their whole or parts for the following acts, while others can consume their contents, leaving only what has been physically left from the operation.

It is generally impossible to have a universal set of operations for the physical world as we have for the information world, so any further inclusion of them into the language may strongly depend on the application areas and concrete system implementation.

EXAMPLES OF WORKING WITH PHYSICAL MATTER

- "5 tons of sand" + "2 tons of sand"—will result in "7 tons of sand".
- N = "200 nails"; M = N * 0.5—will result in "100 nails" in the variable M, with the rest, "100 nails" too, left in the variable N.
- "200 nails" * 2—will fail if the system implementation cannot replicate the physical objects, here nails.
- NboardPlank = ("board", "plank") % "20 nails"—will connect a board and a plank into one board–plank object by "20 nails", assigning it to nodal variable NboardPlank.

- NboardPlank |= "nails"—will separate the compound object kept in nodal variable NboardPlank by removing all nails connecting its two parts. The result as a vector with two scalars ("board", "plank") will be assigned back to the nodal variable NboardPlank, instead of its former single-scalar value.
- ("bar1", "bar2", 'angle', 30) ? weld—will connect the two bars "bar1" and "bar2" by an external welding procedure weld, with the angle of 30 degrees between the bars. As can be seen, the parameters for this procedure are both physical objects and information.

3.10 CONCLUSIONS

A universal language has been described for parallel and distributed creation, management, and simulation of any distributed and dynamic worlds. It navigates distributed worlds by self-evolving spatial recursive programs, called waves, which can set up any control over the worlds as well as create these worlds from scratch and change their structure and behavior at run time.

The peculiar recursive definition of waves allows for a powerful and compact expression of arbitrary complex space navigation, data processing, and control operations, which can be performed in a fully distributed and highly parallel mode.

4

DISTRIBUTED WAVE-WP INTERPRETATION IN DYNAMIC ENVIRONMENTS

We will consider here the main peculiarities of the distributed interpretation of WAVE-WP language by dynamic networks of doers, the latter comprising the language interpreters embedded in communicating vehicles of any technical or biological origin.

The solutions discussed can allow spatial WAVE-WP scenarios to evolve in various environments and set up advanced command and control over other systems, with automatic decision making at any levels. Different interaction and cooperation patterns of the distributed interpretation will be considered, involving distributed processing and transference of both information and physical matter or physical objects in space, as well as between doers.

4.1 DOERS AND THEIR NETWORKS

The WAVE-WP language is executed by a network of software or hardware interpreters, which may have their own physical bodies (as vehicles of any technical or biological origin). These bodies can allow the interpreters not only process information but also move in physical space and perform physical jobs.

Ruling Distributed Dynamic Worlds, by Peter S. Sapaty
ISBN 0-471-65575-9 Copyright © 2005 John Wiley & Sons, Inc.

With the bodies, they can carry, process, deliver, and pass physical matter to other interpreters and communicate with them mechanically, electrically, and/or electronically. The vehicles can also effectively protect the interpreters, allowing them to function properly and, possibly, survive in different environments.

So, in general, we will consider a doer as a combination and integration of the WAVE-WP interpreter and a vehicle. The latter may have a very broad meaning, ranging from separate, evident, and active physical bodies to functional parts of other mechanical, electronic, or social systems, which may themselves have distributed nature. Internet host, humanoid robot, unmanned aerial vehicle, computerized soldier, kitchen stove or a car with a computer, and Hollywood or a United Nations committee may be considered as vehicles—and doers, if supplied with the WAVE-WP interpreter.

Some general and symbolic picture of the execution of a high-level WAVE-WP scenario in an integration of a stationary computer network with mobile manned or unmanned doers, engaged on the demand of the scenario, may look as shown in Figure 4.1.

This WAVE-WP scenario may start from any doer, stationary or mobile, and during its evolution in the united physical–virtual world it may spread to other stationary or mobile doers with WAVE-WP interpreters (or WI) located in the needed area. The scenario interpretation may also request new doers staying elsewhere (free ones or already engaged but with, say, underused capacity), bringing them to the operational theater if needed.

Parts of the local or global scenario may subsequently enter these doers and evolve there, cooperating seamlessly with other parts. If the doers selected are working within other system ideologies or technologies, the scenario may enter them together with its own WI code, setting the needed interpretation environment and evolving within it after an automatic installment of the interpreter.

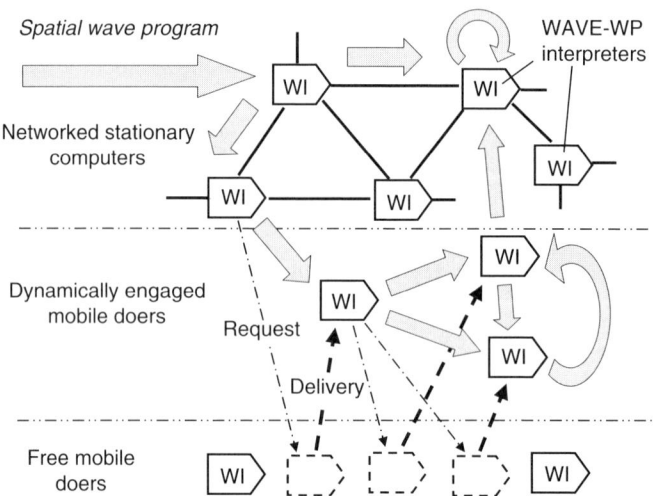

Figure 4.1. Unified organization of stationary and mobile systems in WAVE-WP.

Taking into account that doers can constantly move and change positions both absolute and relative to each other, the communication network between them may have a dynamic and open structure.

4.2 WAVE-WP INTERPRETER ARCHITECTURE

In our previous book (Sapaty, 1999a), we discussed the distributed interpreter architecture in detail in relation to the predecessor language, WAVE. So here we will highlight only basic features of the distributed interpretation of the new language, WAVE-WP, which differ radically from the existing WAVE interpreter or extend it.

These especially relate to the work with physical, execution, and combined worlds, physical matter delivery and processing, and essentially higher level and extended functionality of the current language as regards its predecessor WAVE.

Also, due to patenting of the new interpreter, many sensitive details of its internal organization have to be omitted in this book.

4.2.1 Main Interpreter Components

The general organization of the interpreter is shown in Figure 4.2, with processors (depicted as boxes), data structures (as ovals), and main functional and operational links between them. In brief, the sense and functionality of different components are as follows.

Control Processor. Coordinates all other activities of the interpreter, forwards the incoming waves to the parser, dynamically creates and manages internal

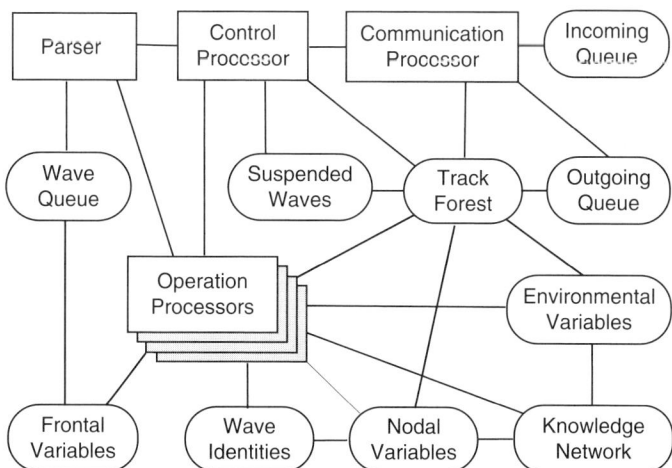

Figure 4.2. General organization of the WAVE-WP interpreter.

interpretation infrastructures (tracks), and executes language rules working on tracks. It makes global optimization and system-level decisions of moving parts of the knowledge network to other interpreters. Keeps further waves in the suspended waves store, linking them with rules in the tracks-based internal infrastructure, and subsequently returning to the waves queue or forwarding to other interpreters.

Parser. Makes syntactic analysis of incoming waves, and if they satisfy the language syntax, puts them into the wave queue. The parser manages wave queue, with parallel execution of waves. It decomposes waves into sequential and parallel moves, provides run time interpretation of expressions, establishing execution order, and also launches subordinate waves. The parser extracts elementary acts in the expressions, supplying them with operands for different operation processors, passing acts to the latter and collecting results. It also forwards suspended waves to the suspended waves store via the control processor.

Operation Processors. These execute elementary acts of the language, including movement in virtual or physical spaces (accompanied by the creation of new track nodes and links), and changing the doer for existing nodes. In order to work with physical matter or physical objects, certain operation processors may need proper hardware or (and) software extensions, which may be part of their own doer or may relate to other doers or systems, while providing global control over these operations. Operation processors may work in parallel with each other, also simultaneously with other modules of the interpreter.

Communication Processor. Manages incoming queue and outgoing queue, exchanging with them (and between them, directly, in transit cases) waves, control and data echoes, and remote results. It may have an external extension (both electronic and mechanical) for physical communication with other interpreters (doers), say, for finding the latter both selectively and in a broadcasting mode in virtual and (or) physical spaces, and passing physical matter or physical objects between doers, as well as for transference of long information files.

Wave Queue. Keeps wave programs that entered this WI, in the first-come–first-served (FCFS) manner, also their further parts into which they split and (or) replicate, as new waves. Moves in waves can be complex expressions, parts of which can be arbitrary waves themselves (due to recursion), executed stepwise or in parallel (also launching subordinate waves and waiting for the decisions from rules and remote results). So the FCFS strategy may often be symbolic, reflecting general priority only, as many waves in the wave queue may be processed simultaneously and may depend on each other.

Suspended Waves. Holds newly formed and suspended branches of the sequential branching rules, also the tails of waves suspended until rules embracing previous parts are completed. From this storage, waves are further forwarded back to the wave queue of the current interpreter directly or via tracks or to the similar queues in other interpreters—via tracks. The order in which waves are kept in this storage does not reflect their engagement order, the latter depending on many circumstances of the execution of distributed scenarios, and especially on rules and tracks.

Track Forest. Keeps fragments of the distributed internal infrastructure related to the current WI, with links to track forests in other interpreters, which altogether form a seamless distributed interpretation infrastructure, which may cover the whole system (and the world, if the system covers it!). It also reflects the history of spreading of waves and accumulates global awareness of the navigated world.

Tracks are used as a dynamically formed *distribution and collection infrastructure*, or *bridges*, by the following waves, also for collecting and merging remote termination states and making decisions by rules by within the echo processes. They help to effectively collect and process in parallel the remote results, optimize an access to the previously visited nodes, as well as to determine and support the lifetime of nodal variables.

Wave Identities. This data structure keeps identities of waves, which allow differently colored waves to make an access in nodes to nodal variables only with the same color, not interfering with the information related to other waves. (The wave's color may be accessed by the environmental variable PAINT from within waves, as described in the previous chapter.)

Nodal Variables. Contains either full contents of nodal variables related to all nodes of the part of the knowledge network residing in the current WI (variables formed by waves that traversed this WI and are still active in the distributed space)—if these contents represent information only and are short enough. Otherwise, the contents of these variables may be a reference to the location of long information files or to physical matter or physical objects, which may reside outside the interpreter and can be managed by external systems.

Frontal Variables. This one contains contents of frontal variables, which are the sole property of waves staying now in the current WI, not nodes. They are brought to the current WI by waves or created while waves were processed in it, and will leave with the waves when they move to other interpreters. Similar to nodal variables, their contents may be either full contents of variables or references to the external full contents (physical matter or physical objects including), and the latter may have to move with the current doer or between doers physically.

Environmental Variables. This unit implements and supports the environmental variables of WAVE-WP, which may be related to individual multiple nodes and individual multiple waves kept in the current WI, and also to the interpreter itself and its underlying hardware, as well as to the external world.

Knowledge Network. Keeps part of the distributed knowledge network that resides in the current interpreter, that is, its nodes and links between them, as well as links to nodes located in other interpreters, forming with the other parts of KN a seamless distributed and integral network space. KN nodes and links may represent the united world described in the previous two chapters, combining and merging features of the physical, virtual, and execution worlds. The link and node contents may represent links and nodes completely or may be references to their outside locations, especially when they represent physical objects or matter.

Incoming Queue and Outgoing Queue. These are temporarily storing different types of messages being exchanged with other WAVE-WP interpreters, located in other doers. The general service mode for these two stores is FCFS, but different types of messages may have different priorities, say, echoed generalized termination states (among which may happen to be `abort`) have higher priority in the incoming queue than the returned remote data or new waves.

The `abort` message in the outgoing queue has an absolute priority, and also messages containing a value to be assigned to remote variables are prior to the new waves to be sent to other interpreters. So the order of records in these two stores may differ from their service order.

4.2.2 Exemplary Interpretation Patterns

To understand a bit more how the WAVE-WP interpreter works, we will consider here some simple patterns of communication and interaction between the main components of the interpreter, as well as data flow between them, which are shown in Figure 4.3.

Interpretation of Expressions. Plain bold arrows show the participation of components in interpretation of expressions within the same interpreter (only main

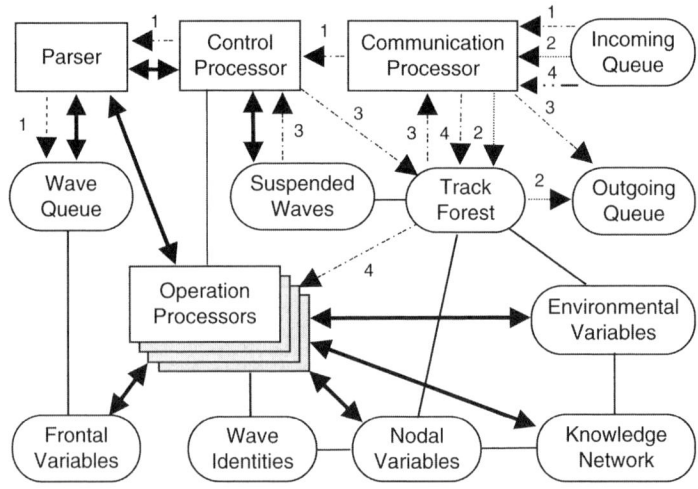

Figure 4.3. Some WAVE-WP interpretation patterns.

WAVE-WP INTERPRETER ARCHITECTURE

functional links are highlighted). It involves the control processor, parser, operation processors, wave queue, suspended waves, nodal variables, frontal variables, environmental variables, as well as the knowledge network, all under the general guidance of the parser.

New Incoming Waves. Dashed arrows (also marked by 1) show the technological path for new incoming waves: incoming queue, communication processor, control processor, parser (making preliminary syntactic analysis), and wave queue, under the general guidance of control processor. From wave queue, these waves will again be used by the parser for a further execution.

Waves Passing Through. Dotted arrows (also marked by 2) show a transit path for waves that are just passing through the current interpreter (without being processed in it) and directed by the tracks out of it to other interpreters: incoming queue, communication processor, track forest, and the outgoing queue—under the general guidance of the communication processor.

Forwarding Suspended Waves. Dash–dotted arrows (also marked by 3) show the path of forwarding suspended waves to other interpreters via the track infrastructure: suspended waves, control processor, track forest, communication processor, and the outgoing queue—under the general guidance of the control processor.

Returning Remote Data. Dash–double-dotted arrows (marked by 4) show the path of returning remote data for further processing in an expression in the current interpreter: incoming queue, communication processor, track forest, and the operation processor—under the guidance of both the parser and the communication processor. In a reverse order and the reverse direction of arrows, the data obtained in one of the operation processors will be passed through tracks and then outside the interpreter, to other interpreters, to be further processed there and/or possibly assigned to remote variables.

4.2.3 Integration of the Interpreter with Other Systems

The interpreter, if to be installed in manned or unmanned mobile platforms, may be integrated with special software or hardware modules being part of these platforms, forming altogether what we call a *doer* in this book. These modules may coordinate continuous motion in space, provide vision, manipulation with physical objects, and the like, as well as communicating with other such platforms using, for example, radio, radar, laser, or sonar channels.

This is shown in Figure 4.4 where MC stands for motor control, VC is for vision control, S represents sensors, C for communication facilities, PMO identifies physical matter or object manipulating or processing tools, and NC is for neurocomputers, which can take responsibility for local run time optimization and control of different operations. The special modules shown in the figure may have external extensions, as specific tools or gadgets, which may be physically attached to doers or can be located and used remotely.

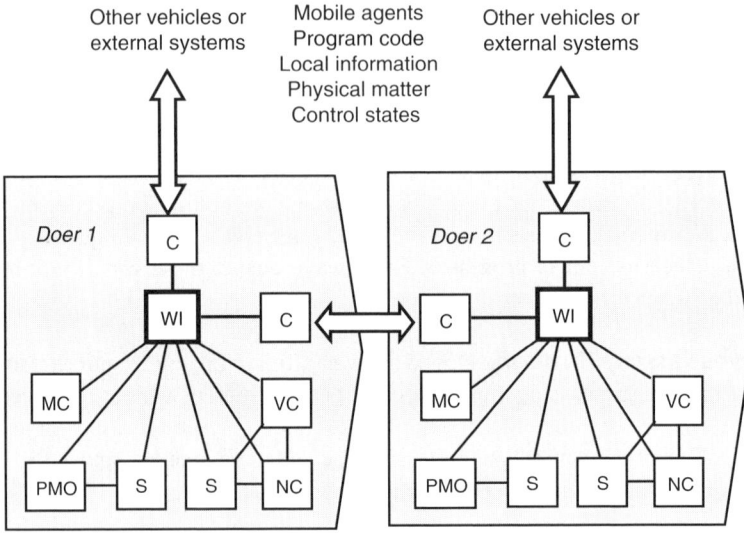

Figure 4.4. Integration of WAVE-WP interpreter with other systems.

The WAVE-WP interpreters establish a higher coordination and control level over such platforms, also providing their ability to work together as one system, with one united albeit physically distributed artificial brain. This brain can seamlessly hold, acquire, and accumulate any distributed knowledge and process it in parallel within the integral mission scenarios written in the spatial language. During the interpretation of WAVE-WP language, different forms of information and physical matter can be exchanged between doers, as shown in Figure 4.4.

4.3 TRACK INFRASTRUCTURE

Scenario-oriented automatic spatial control is based on *dynamic tracks*. These serve as a hierarchical machine coordinating the spread of waves, assessing distributed situations, making autonomous decisions (involving rules), transmitting new waves to the locations reached by the previous waves. Tracks also collect and return remote results, support and determine lifetime of distributed variables, and terminate multiple remote processes. Tracks self-optimize at run time, forming intelligent internal infrastructures hidden from applications.

4.3.1 Forward and Backward Operations

Track nodes and links are created when waves, accompanied by frontal variables, navigate in the united world and pass knowledge network nodes, as shown in Figure 4.5. Created nodal variables are linked with track nodes relating to the passed KN nodes.

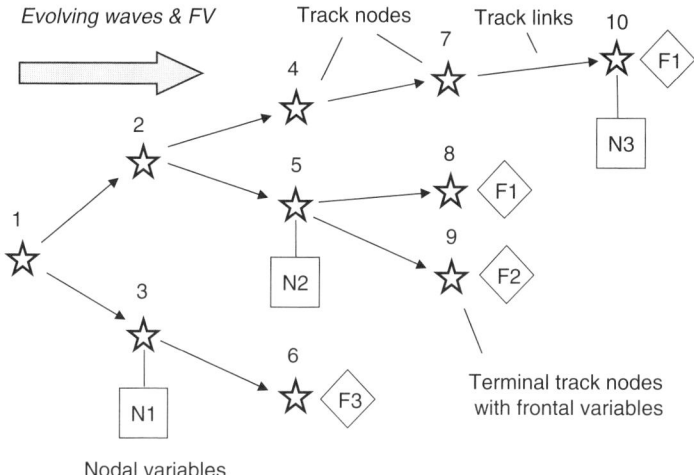

Figure 4.5. Forward processes with run time creation of tracks.

The existence of these track nodes determines the lifetime of the nodal variables. The fringe track nodes correspond to the final nodes reached by the wave (i.e., its SNR), and frontal variables carried with the waves, after the latter expire, will be finally linked with these fringe track nodes, as shown in Figure 4.5.

In a parallel backward processes, starting from the fringe nodes with certain termination states and finally reached (or obtained) data values of waves, the track tree echoes and merges states and data, receiving the final echo results in the root node; see Figure 4.6. These echo results may be further used in the distributed control governed by rules or in wave expressions as the obtained remote data for further processing.

Local optimization of the track network also takes place in the backward process, where intermediate nodes not linked to nodal variables may be removed and bypassed (such as nodes 2, 4, and 7 in Fig. 4.5) with direct links set up between the predecessors and successors of these removed nodes, as shown in Figure 4.6 (namely, $1 \rightarrow 10$ and $1 \rightarrow 5$).

Any further wave code (say, suspended by rules), starting from the root of the created track tree, will replicate in track nodes and propagate down the track tree, as shown in Figure 4.7. Upon reaching the fringe nodes of the track tree, the further waves will integrate with frontal variables left there by the previous wave, forming new full-scale wave programs, which will develop in the united world from the SNR of the previous wave.

4.3.2 Optimization of the Track Network

As already shown, the track network is dynamically optimized to be the simplest and at the same time capable of fulfilling its functionality. This makes the root node maximum

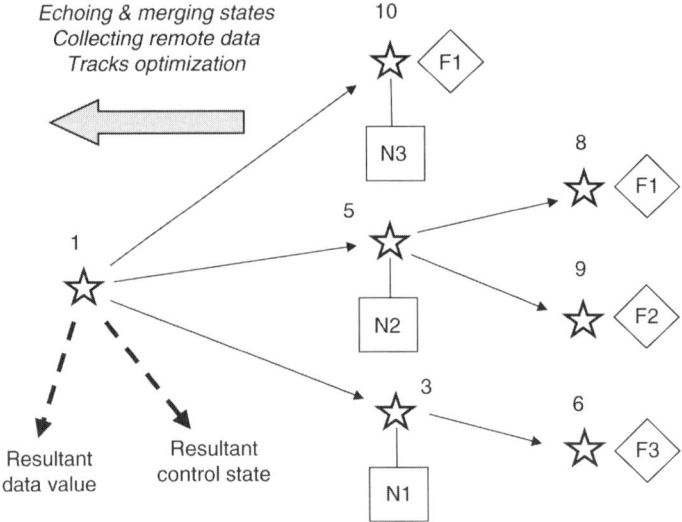

Figure 4.6. Backward track processes.

close to the fringe nodes, speeding up the transference through tracks of further waves, and also saves the track nodes responsible for nodal variables created in KN nodes by the previous wave.

The network track infrastructure can be made more advanced and better manageable by establishing bypass links from the root node to the fringe nodes, avoiding nodes bound to nodal variables. Also, horizontal links between the fringe nodes and

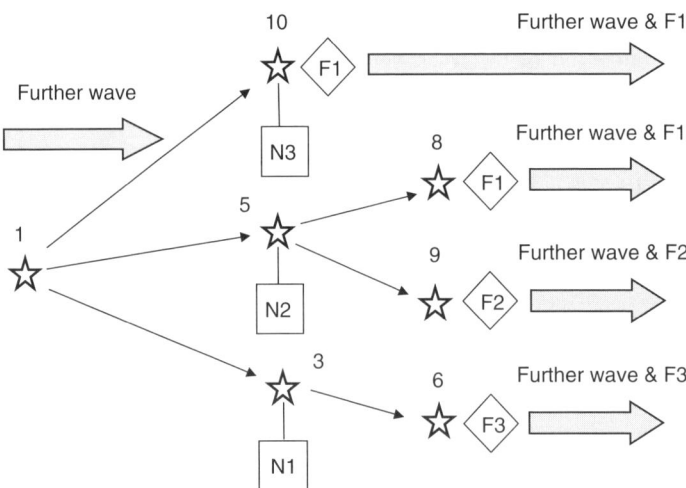

Figure 4.7. Spreading further waves, integration with frontal variables.

TRACK INFRASTRUCTURE

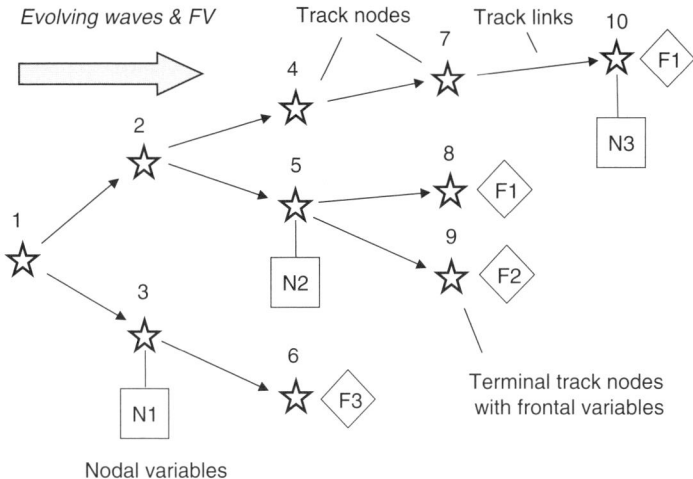

Figure 4.5. Forward processes with run time creation of tracks.

The existence of these track nodes determines the lifetime of the nodal variables. The fringe track nodes correspond to the final nodes reached by the wave (i.e., its SNR), and frontal variables carried with the waves, after the latter expire, will be finally linked with these fringe track nodes, as shown in Figure 4.5.

In a parallel backward processes, starting from the fringe nodes with certain termination states and finally reached (or obtained) data values of waves, the track tree echoes and merges states and data, receiving the final echo results in the root node; see Figure 4.6. These echo results may be further used in the distributed control governed by rules or in wave expressions as the obtained remote data for further processing.

Local optimization of the track network also takes place in the backward process, where intermediate nodes not linked to nodal variables may be removed and bypassed (such as nodes 2, 4, and 7 in Fig. 4.5) with direct links set up between the predecessors and successors of these removed nodes, as shown in Figure 4.6 (namely, $1 \rightarrow 10$ and $1 \rightarrow 5$).

Any further wave code (say, suspended by rules), starting from the root of the created track tree, will replicate in track nodes and propagate down the track tree, as shown in Figure 4.7. Upon reaching the fringe nodes of the track tree, the further waves will integrate with frontal variables left there by the previous wave, forming new full-scale wave programs, which will develop in the united world from the SNR of the previous wave.

4.3.2 Optimization of the Track Network

As already shown, the track network is dynamically optimized to be the simplest and at the same time capable of fulfilling its functionality. This makes the root node maximum

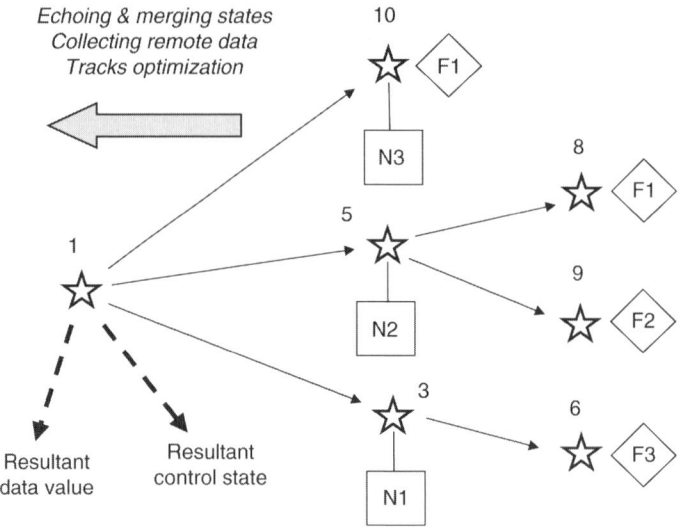

Figure 4.6. Backward track processes.

close to the fringe nodes, speeding up the transference through tracks of further waves, and also saves the track nodes responsible for nodal variables created in KN nodes by the previous wave.

The network track infrastructure can be made more advanced and better manageable by establishing bypass links from the root node to the fringe nodes, avoiding nodes bound to nodal variables. Also, horizontal links between the fringe nodes and

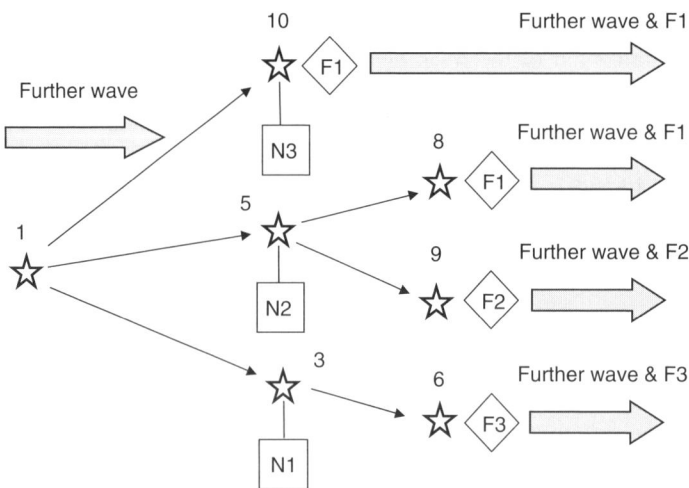

Figure 4.7. Spreading further waves, integration with frontal variables.

ELEMENTARY OPERATIONS INVOLVING MULTIPLE DOERS

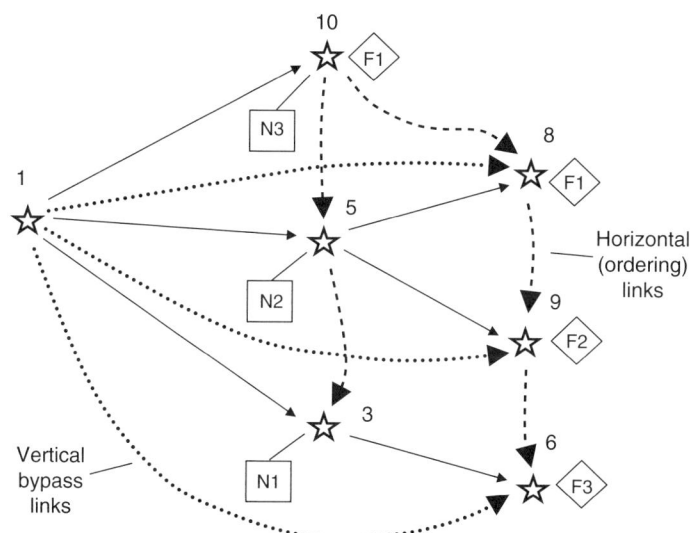

Figure 4.8. More advanced track system with bypass and horizontal links.

links between internal nodes that hold nodal variables, can bed set up, as shown in Figure 4.8.

The new links may be oriented ones, reflecting the order of the world navigation and track infrastructure creation. This recorded order may improve further navigation through tracks, say, accessing fringe nodes in the order they were created, which is essential for sequential branching rules of WAVE-WP language. The additional links may also speed up the search for proper KN nodes previously passed by waves, thus improving distributed and global awareness of the navigated world.

4.4 ELEMENTARY OPERATIONS INVOLVING MULTIPLE DOERS

To hint at what is actually going on at the cooperative interpretation level of the language, we will consider spatial execution of some elementary operations by the distributed interpreter of WAVE-WP. The legend helping to explain the pictures is shown in Figure 4.9.

4.4.1 Local Operations in Doers

We will start with elementary operations performed entirely within single doers. Assume we need data processing in a node (let it have name a), say, an assignment to nodal variable N of the sum of frontal variable F and value 10, as follows:

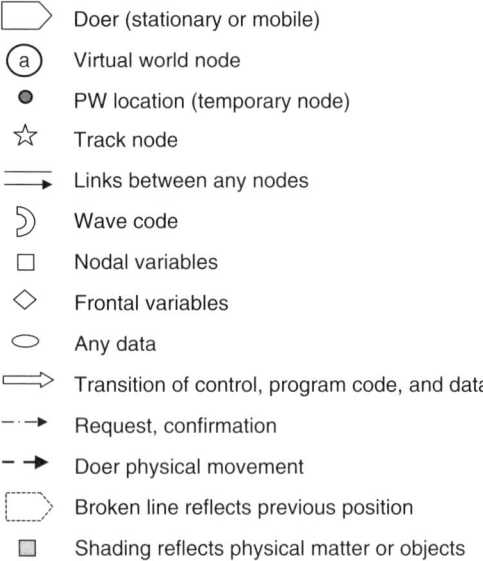

Figure 4.9. Main symbols used and their meaning.

```
N = F + 10; wave
```

Variable `N` associated with node `a` will be assigned the new value, the control will continue to reside in this node, and the remaining `wave` will be applied from the same node a, as shown in Figure 4.10.

In another example, let us (starting again from node a) create a new virtual node, say, `b` with link named `p` to it from node `a`. For this, the WAVE-WP code will be as follows:

```
create(p # b); wave
```

where `wave` represents the rest of the wave program too.

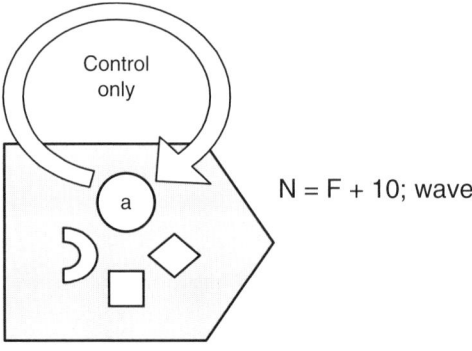

Figure 4.10. Assignment in a node.

ELEMENTARY OPERATIONS INVOLVING MULTIPLE DOERS 111

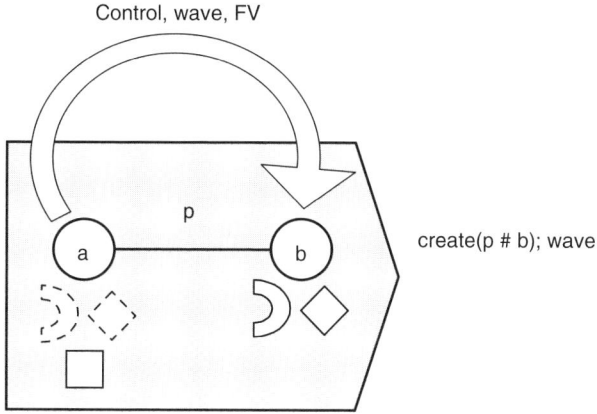

Figure 4.11. Creating a new node in the same doer.

The distributed interpreter can decide whether this extension of the virtual space should be done within the same doer or should be placed into another doer. Placing the new node with new link into the same doer is shown in Figure 4.11, where control, the rest of the program, and associated with it frontal variables will be relinked to the new node within the same interpreter (and therefore doer).

4.4.2 Creating a New Virtual Node in Another Doer

If for the previous example the system decides to put the node into another doer, say, in order to optimize the information distribution between doers, we will have the situation shown in Figure 4.12. The two nodes will be located now in different doers (doer.1 and doer.2), and the link between them will be inherited as a new semantic relation between the two doers themselves.

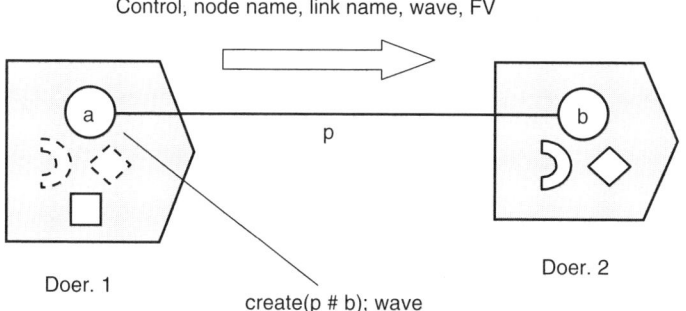

Figure 4.12. Creating a new node in another doer.

We can also instruct the interpretation system to create the new node in the explicitly named doer (let it again be doer.2), as follows:

```
create(p # b_doer.2); wave
```

The creation of the new node in another, possibly remote, doer will be accompanied by transference of the rest of the program together with frontal variables to this new doer, also leaving track control with the previous doer (the latter is implied but not shown in Fig. 4.12, it will also be implied in the following operations). This transference (which also involves new node and link names) may be done electronically through a direct communication channel (or communication network, possibly, via other doers), if the frontal variables associated with the wave program keep information only.

Transference of Physical Matter Between Doers. The previous example may be complicated if the creation of a new node in another doer is accompanied by the transference of physical matter to this doer, say, when the matter (or object) is kept in a frontal variable associated with the wave program, which has to move to the new node, as follows:

```
F = "tree"; create(p # b_doer.2); wave
```

To solve this problem, one of the doers (or both) may have to move in the physical world and make the transference of a physical object in a direct contact, as shown in Figure 4.13 for doer.1 moving.

Another solution may be when doer.2 is moving on a request from doer.1 (this request should contain exact coordinates of doer.1 in the physical world), as shown in Figure 4.14.

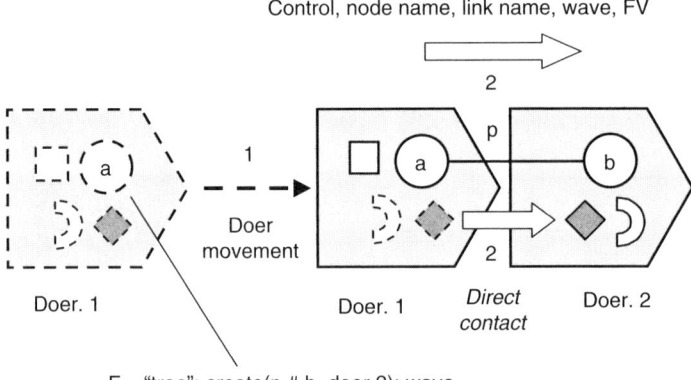

Figure 4.13. Creating a node in another doer and passing the physical object, with doer.1 moving for a contact.

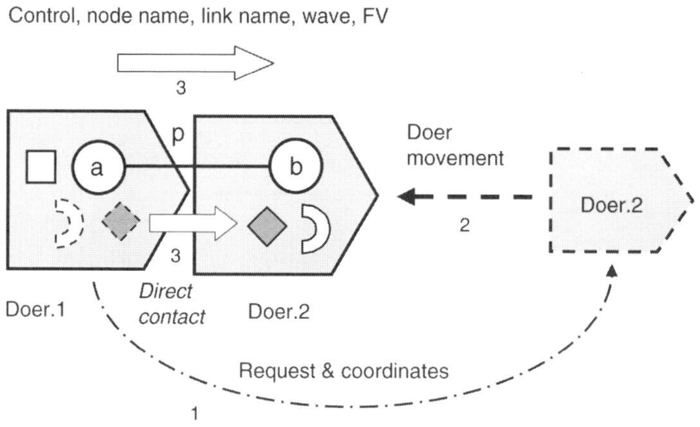

ELEMENTARY OPERATIONS INVOLVING MULTIPLE DOERS

Figure 4.14. Another solution where `doer.2` is moving to a direct contact.

Both doers can also move to some meeting point, say, to complete this combined operation of node creation and physical object transference quicker.

Involving a Third Doer. If both doers cannot move physically to meet, say, being fixed in certain PW locations or are responsible for PW-bound nodes (these and other options are provided by WAVE-WP language), another, third, doer (let it be `doer.3`) may be requested to make the transference of physical matter between them. The control and rest of the program can be passed to (remote) `doer.2` either electronically, straightforwardly, as in Figure 4.12, or within a direct electrical connection first between `doer.1` and `doer.3` and then between `doer.3` and `doer.2`, along with transference of the physical matter or object (PMO). In Figure 4.15, the electronic transmission between `doer.1` and `doer.2` is shown as performed after the transmission of physical matter between them via `doer.3` (numbers at arrows reflect the order of operations).

4.4.3 Moving into a New Physical Location

Another class of interpreter operations relates to visiting new positions in the physical world, where a hop has been met in the wave program that leads to a certain PW location given by is coordinates, as follows:

```
direct # x44 y37.8; wave
```

Here, a variety of options may be available too.

Current Doer Moving Itself. First, the same doer, interpreting the current wave code, can physically move into the new place in PW, as shown in Figure 4.16 (where the above program is applied in the current doer's PW position, let it be `x12 y22`).

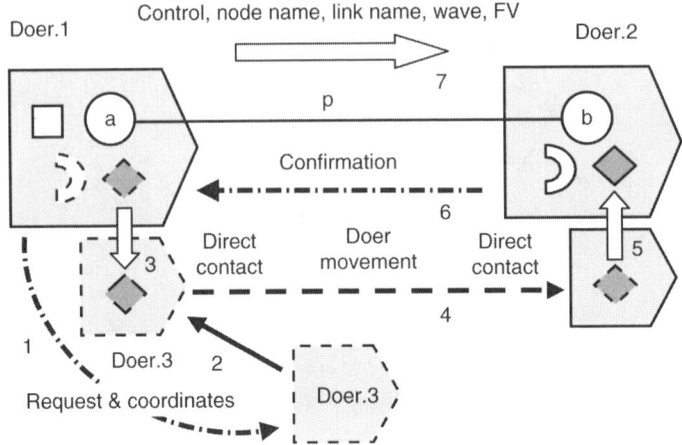

Figure 4.15. Involvement of a third doer for the transference of PMO only.

The new temporary node x44 y25 will be entered by the rest of wave (i.e., wave) together with frontal variables (which will be withdrawn from the previous node). The fate of the previous node x12 y22 with existing nodal variables in it will be determined by the presence of other activities in it. This node will cease to exist (together with nodal variables) if no other activities are linked with it, following the semantics of temporary PW nodes in WAVE-WP.

Requesting Another Doer to Move. Second, another doer can be requested to move into the new PW location. And to this doer, the rest of the wave program (along with program control and frontal variables, if the latter keep information only) can be electronically transferred (before, during, or after reaching the new PW location). A possible order of these operations is shown in Figure 4.17, reflected by the numbers at the arrows.

A virtual copy of the destination node x44 y25 can first be set up in doer.2 along with the rest of the wave program and its frontal variables, which, upon reaching the actual destination in the physical world, will be converted into the physical node.

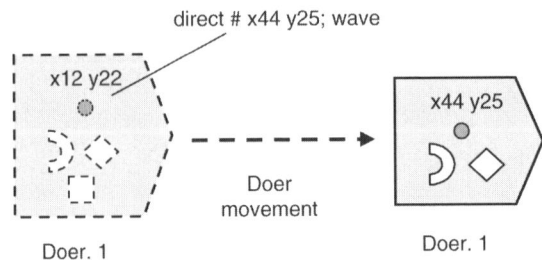

Figure 4.16. Reaching a new physical world location by the same doer.

MORE COMPLEX SPATIAL OPERATIONS 115

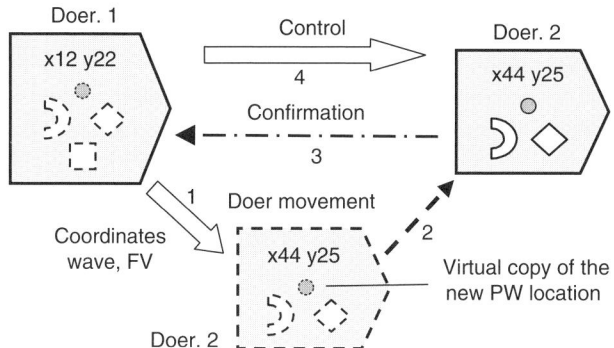

Figure 4.17. Using another doer for reaching the new location.

Other Variants of Movement. Other variants can be possible too, say, first moving an unloaded doer 2 to the x44 y25 location and then electronically transferring into the newly created node the wave code, frontal variables, and program control. Using the other doer for reaching the needed PW destination can also be stated explicitly in the wave program, as follows:

```
direct # x44 y25_doer.2; wave
```

If physical matter is associated with the current node in frontal variables, other options may be needed. For example, doer.1 or doer.2 (or both) should first move in PW to make physical contact and transfer the matter, before doer.2 moving to the new physical location. Or a third doer should be engaged for the transference of matter between doer.1 and doer.2, as in the previous section, when the new virtual node was created.

It is also possible in WAVE-WP to change the current position in physical world without considering it as a new node—by just assigning the new physical coordinates to the environmental variable WHERE, as follows:

```
WHERE = x44 y25; wave
```

Possible implementation solutions for this case will be similar to the ones considered above.

4.5 MORE COMPLEX SPATIAL OPERATIONS

In the previous sections we considered only the simplest cases of distributed interpretation of the WAVE-WP operations. Some more complex examples are discussed in this section.

4.5.1 Moving Data Through Tracks

A variety of distributed processing and control operations use the distributed track infrastructure, and transferring via tracks of both information and physical matter are most typical of these operations, allowing us to coordinate and manage the distributed worlds seamlessly.

Transference of a Large Volume of Information. In Figure 4.18, a situation is shown where a certain (presumably large) amount of data is needed to be transferred from a source to a destination, where the latter can be found by traversing track links first, starting from the source doer. The found destination doer address is then returned to the source doer, which organizes direct transference of the large data volume to the found destination doer, using for this data transmission, say, special broadband channels.

Transference of Physical Matter. If the data to be transferred is physical matter or physical objects, the situation may be as shown in Figure 4.19, where, after finding the needed destination doer via tracks, its exact physical coordinates are returned to the source doer, rather than the electronic address as in the previous case.

Using the found destination doer coordinates, the source doer can then move physically to reach the destination doer, transferring the physical matter to it in a direct contact. Another solution may be the transference via the track infrastructure of the physical address of the source doer to the destination doer, with the latter subsequently moving itself to the source doer to get the physical matter in a direct contact. Using a third, service, doer for transferring between the source and destination doers (to whom both the source doer coordinates and the found destination doer coordinates are transferred first) may be another option.

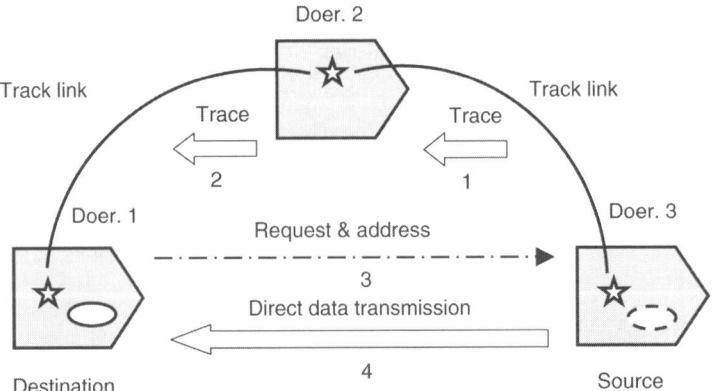

Figure 4.18. Transference of large data volumes using distributed tracks for finding the destination doer.

MORE COMPLEX SPATIAL OPERATIONS

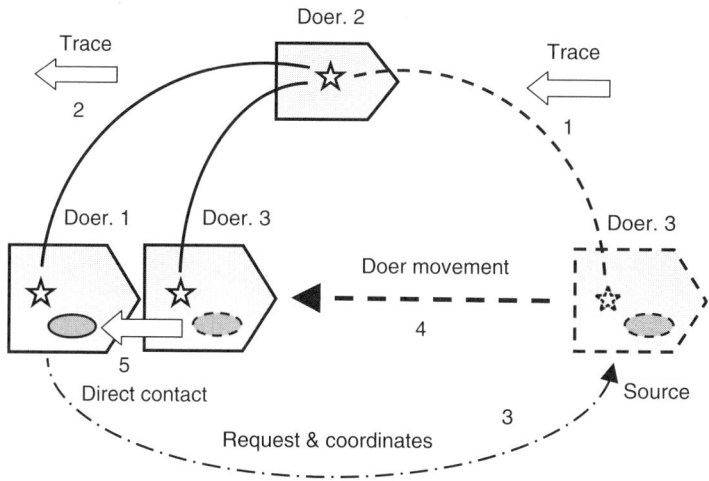

Figure 4.19. Transference of physical matter via tracks.

4.5.2 Migration of Knowledge Networks Between Doers

WAVE-WP allows us to distribute virtual worlds arbitrarily between doers, and these worlds may be active while being inhabited by multiple migrating processes. Different local wave processes and data, including physical matter, may be associated with the nodes at each moment of time. Nodes of the virtual and combined worlds can migrate between doers at run time under different circumstances.

Imagine that for some reason node b in doer.1 (say, doer.1 proved to be overloaded or not safe enough) has to be transferred to another doer, say, doer.2, together with all wave programs, local information, and physical matter currently associated with it. This change of doer should preserve the existing logical links of node b with other nodes, which may be now in other doers, and also all the active processes associated with node b, should these continue operation in b after it is transference to doer.2.

In Figure 4.20, the transference of node b from doer.1 to doer.2 is depicted, which involves service doer (doer.4) for transportation of physical matter associated with the node b (kept in both nodal and frontal variables).

Dashed links reflect the previous position of node b in doer.1, its links p and q with nodes a and c in doer.1 and link s with node d in doer.3, as well as its associated waves and nodal and frontal variables in doer.1.

This node transference operation can also be ordered explicitly from the wave program, when some wave, staying in node b, assigns doer.2 to the environmental variable DOER:

```
DOER = doer2; wave
```

More complex situations can occur when united virtual–physical nodes will have to be transferred to other doers, where the destination doers will also have to move to the physical locations associated with these nodes.

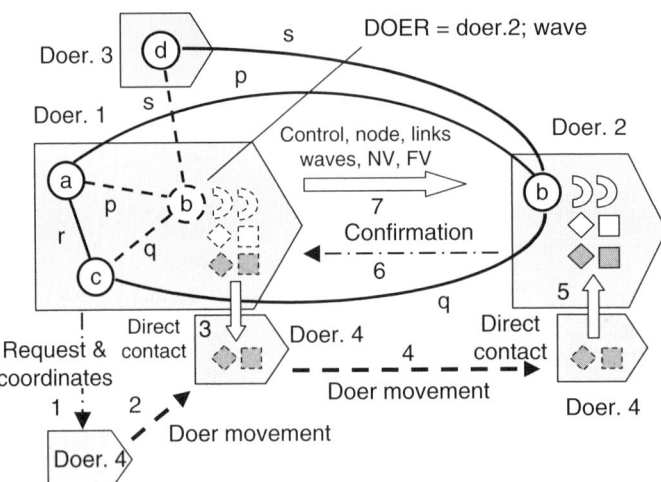

Figure 4.20. Moving a node to another doer using a service doer for the transference of physical matter.

We have considered only a few situations of transference of information, program code, and physical matter between doers, as well as physical movement of doers, triggered by elementary operations of WAVE-WP language or by internal self-optimization of the interpreter. In distributed interpretation of WAVE-WP, many such cases may need to be considered together, along with other, more complex spatial operations.

4.6 OTHER DISTRIBUTED INTERPRETATION ISSUES

4.6.1 Mapping Strategies

Different strategies of invocation of new doers during the evolution of spatial WAVE-WP scenarios in the united physical–virtual world can be used. The simplest one may be their delivery to the needed points from a sort of central depot. Another strategy may resemble calling taxicabs, when free doers are delivered from the nearest locations (put there initially or released by other waves).

The solution may also be based on run time analysis of waves and prediction of their forthcoming needs in mobile hardware, the latter propagating with the waves as spare horses and engaged (loaded with waves and frontal variables) when needed. Combinations of these strategies may be useful too.

4.6.2 Dealing with Shortage of Vehicles

Optimization of Movement in Space. Physical movement in space may be optimized. An interpreter need not necessarily physically stay all the time in PW

OTHER DISTRIBUTED INTERPRETATION ISSUES 119

locations corresponding to the PW-bound nodes it hosts, as many operations defined in these nodes may be information processing only, not using physical world parameters associated with the nodes. So these operations can well be performed wherever the doer is staying physically at this moment. A return to the PW locations of these nodes may, however, be vital if PW qualities should be accessed in these nodes, a physical job to be done in the corresponding PW location, or a physical hop to other locations from this node is to be performed.

A demand to a return to the PW node for making a planned hop from it to another PW node may, however, not be absolute too. The hops to nodes from certain locations in space may often be substituted by hops to the same destinations from a current position of the interpreter, of course, if conditions (say, a terrain) permit this.

Parallel Solution for Space Navigation. As an example of using this tactic, let us consider the following symbolic program describing parallel operations in the physical world:

```
#x1 y3; w1;
(# x3 y5; w2), (# x4 y3; w3), (# x3 y1; w4);
# dx3 dy0; w5
```

In this program (its maximum parallel execution by three doers is shown in Fig. 4.21a), it is assumed that w1 to w5 are programs operating in the same nodes only, without moving to other nodes and leaving final results in these nodes. After a hop to the starting point x1 y3 (marked as node 1 in Fig. 4.21a), wave w1 is executed

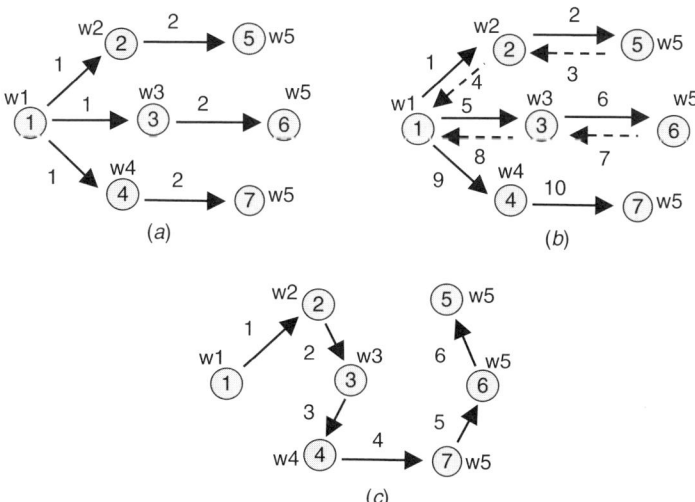

Figure 4.21. Examples of execution of a wave by different number of mobile doers: (a) maximum parallel by three doers, (b) by a single doer performing all scheduled hops, and (c) by a single doer optimized, with changed hops.

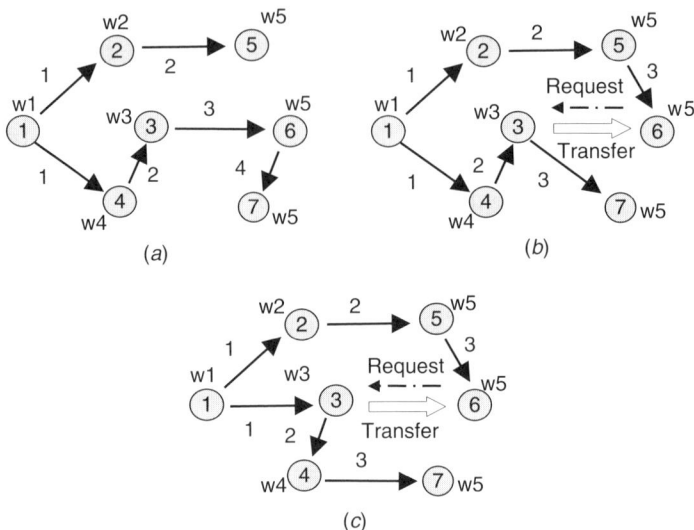

Figure 4.22. Optimized solution by two doers: (a) without communication, (b) with synchronization and communication, and (c) further optimized.

in this node. Then hops to locations x3 y5, x4 y3, and x3 y1 (marked, correspondingly, as nodes 2, 3, and 4) are executed in parallel, with waves, respectively, w2, w3, and w4, operating in the reached destinations, in parallel too.

After a nonsynchronized termination of the latter waves, from nodes 2, 3, and 4 a similar hop in space by coordinate shifts dx3 dy0 will be performed in parallel, reaching the new locations marked as 5, 6, and 7 in Figure 4.21a. The same copies of wave w5 will operate independently in these final nodes. Waves w1 to w5 can perform any data-processing jobs and operate on both nodal and frontal variables, where frontal variables can be inherited by the subsequent waves (with absolutely needed ordering of waves related to their maximum parallel execution shown in Fig. 4.21a, by numbers at arrows).

Assuming that we are already staying in starting node 1, the whole task, as is clear, can be performed in two symbolic (parallel and asynchronous) steps only, for which three doers, operating simultaneously, can be sufficient.

Sequential Solution. If we have only a limited number of vehicles (i.e., two or one), in order to support the program scenario and execute the hops in PW exactly as they are originally defined (and between the exact locations), the doers will need to "sequentialize" some or all jobs (here hops and waves), and also to make returns to the same locations to perform the defined hops.

A possible fully sequential solution of the task above by a single vehicle is shown in Figure 4.21b, where 10 steps will be needed (some steps will be return hops into the same nodes, using the already passed routes in the reverse direction, which is expressed by dashed arrows). In Figure 4.21c, one of the possible solutions is shown how a single vehicle can perform the whole task (obeying the original minimal ordering of waves shown in Fig. 4.21a) in six steps only, if it is possible to substitute original hops by direct hops to the same locations from other points (of course, if this is possible in the physical space, say, terrain permitting).

Possible Solutions by Two Doers. A possible optimal solution with changing hops performed by two doers operating fully independently, without communication with each other, is shown in Figure 4.22a, having a maximum of four steps.

However, if to allow doers communicate and transfer contents of possible frontal variables flowing between waves $w3$ and $w5$, on a request of the doer executing one of the copies of $w5$, the whole job can be done by two doers in three steps only, as shown in Figure 4.22b. This schedule can be further improved if to execute $w3$ as soon as possible and reduce to minimum possible waiting time for $w5$, as shown in Figure 4.22c.

4.7 CONCLUSIONS

The described key principles of the distributed interpretation of WAVE-WP in both virtual and physical worlds may provide high integrity and universality of distributed systems operating within this ideology and technology. The technology can settle on top of other technologies and hardware or software systems as an intelligent management and coordination layer.

Due to WAVE-WP, these systems can acquire a unique capability of behaving as a highly organized whole, effectively pursuing local and global goals and surviving in dynamic and hostile environments despite indiscriminate failures and losses of separate components and communications between them.

5
SPATIAL PROGRAMMING IN WAVE-WP

The chapter contains examples of the practical use of the WAVE-WP language in a variety of situations, ranging from sequential or parallel programming in traditional computers and computer networks to expressing higher level abstractions and cooperative actions in the united physical–virtual space and time continuums, with possible implementation in advanced computer networks and multiple mobile robots. Included are elements of the new integral, agent-free methodology of distributed and parallel programming of dynamic systems in a spatial pattern-matching mode, provided by the WAVE-WP model.

5.1 TRADITIONAL SEQUENTIAL AND PARALLEL PROGRAMMING

5.1.1 Programming in a Single Doer

We start from using WAVE for programming in a single doer, as a traditional computer. This programming can be both sequential and concurrent or parallel.

Elementary Example. Let us consider first an elementary program consisting of two statements (or moves, in the terminology of WAVE-WP) separated by a semicolon:

```
N = 5 + 3; USER = N
```

Put into a single doer (see Fig. 5.1), it works as a usual sequential program, where the first move finds the sum of the two constants and assigns the result to a variable N, and the second move outputs the new content of N by assigning it to the environmental variable USER.

Each move returns control to the same node, which is the doer itself. (The doer node here is the only node belonging to the set of nodes reached, or SNR, produced by each of the two statements.)

Sequential Vector Summation. As more complex, let us consider the program that first defines a vector of four elements, assigning it to a variable N and then sums all elements of N stepwise, sequentially, in a cyclic mode, outputting the sum obtained, as follows:

```
N = (10, 20, 30, 40); Number = 4;
repeat (Ni += 1; Ni <= Number; Nsum += N : Ni);
USER = Nsum
```

Variables (all nodal) Number, Ni, and Nsum are used to keep, correspondingly, the total number of elements of the vector, index of the current element of the vector, and the accumulated sum of elements. The repetitive part of the program increments the index Ni, compares its new value with Number, and adds to Nsum the new, indexed, element of N. It will continue working until the filter act <= returns state thru, otherwise the rest of wave, which is an output of Nsum (i.e., assignment of it to variable USER), will be executed.

To show some flexibility of the language, we may rewrite the program above for the case where the number of elements of the vector N is not known in advance, and the cyclic part of the program works as long as the elements retrieved from vector N by the growing index Ni are not empty (i.e., not nil), as follows:

```
N = (10, 20, 30, 40);
repeat(Ni += 1; N : Ni != nil; Nsum += N : Ni);
USER = Nsum
```

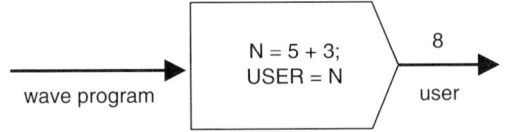

Figure 5.1. Elementary programming in a single doer.

TRADITIONAL SEQUENTIAL AND PARALLEL PROGRAMMING 125

Even more flexible may be the solution where the first element of the vector N is always added to Nsum, after which it is removed from the vector (formally substituted by nil); and the cycle is repeating unless the whole vector N becomes empty (of course, if N is not needed for subsequent computations), as follows:

```
N = (10, 20, 30, 40);
repeat(N != nil; Nsum += N : 1; N : 1 = nil);
USER = Nsum
```

Parallel Vector Summation. Summing elements of the vector may also be done in parallel, by the following program, where just naming a vector provides an independent access to all its elements for a subsequent move, the latter being replicated for each vector element, with all copies working concurrently with the common variable Nsum:

```
N = (10, 20, 30, 40);
sequence((N; Nsum += VALUE), USER = Nsum)
```

An access to individual elements of the vector after the replication of control is provided by the environmental variable VALUE, and the rule sequence is used to separate multiple control of the simultaneous access to all vector elements, from a single control for printing the resultant value in variable Nsum.

The overall control in the program above may be simplified by using the echo rule none, always leaving a single control point in the node that started the rule (here the doer itself, as the only node), regardless of the SNR obtained by the wave it embraces and also regardless of its success or failure, as follows:

```
N - (10, 20, 30, 40);
none(N; Nsum += VALUE); USER = Nsum
```

Using square brackets as an allowed equivalent of none, this program will be even more compact:

```
N = (10, 20, 30, 40);
[N; Nsum += VALUE]; USER = Nsum
```

And, finally, a special echo rule sum can be used to obtain the sum of all values associated with the SNR of the ruled wave—the latter formed by virtual copies of the current node (i.e., the doer), each associated with a separate element of the vector N, as follows:

```
N = (10, 20, 30, 40);
USER = sum(N)
```

If operating over the vector elements directly, without keeping them in a variable, we may get the maximum compact (as well as parallel) program, as follows:

```
USER = sum(10, 20, 30, 40)
```

The echo rule `sum` may provide the most efficient implementation of the vector summation process, as it can be performed entirely on the internal system level.

5.1.2 Programming in Multiple Doers

We will consider examples of parallel and distributed programming in a group of four doers (named `doer.1` to `doer.4`), using the same vector summation example.

Distribution of Information Between Doers. First, we will distribute the vector N between the doers, where each vector element should be located in a separate doer, in the variables each named N too. Parallel distribution of the vector between the doers (defined by the following program) may start from any doer, which may be some other doer or any one of the named four. The following program will do this distribution job:

```
(direct # doer.1; N = 10),
(direct # doer.2; N = 20),
(direct # doer.3; N = 30),
(direct # doer.4; N = 40)
```

This program will split into four independent programs (say, wave1 to wave4), each holding a certain element of the vector and assigning it to the variable N in a separate doer after a hop into it, as shown in Figure 5.2.

This program can be optimized, with only one, common, assignment present, which will be replicated for each doer:

```
(direct # doer.1; 10),
(direct # doer.2; 20),
(direct # doer.3; 30),
(direct # doer.4; 40); N = VALUE
```

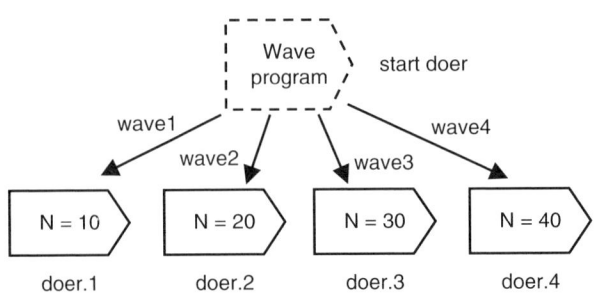

Figure 5.2. Distribution of the vector between four doers.

Distributed Summation Using Migrating Control. There may be a variety of ways of finding the sum of these four elements distributed between the doers. A straightforward one may be stepwise, sequential moving between the doers, without any central control, and adding the values of the elements they hold to a frontal variable Fsum moving between doers together with the mobile control, as follows:

```
direct # doer.1; Fsum += N;
direct # doer.2; Fsum += N;
direct # doer.3; Fsum += N;
direct # doer.4; Fsum += N;
USER = Fsum
```

This sequential process with migrating control is depicted in Figure 5.3, where wave1 to wave4 identify different stages of the program above which gradually reduces its size, losing worked parts after performing the summing operation in each doer and moving to another doer. The resultant content of the moving variable Fsum is finally sent to a user.

Cyclic Solution of the Distributed Summation. The program above can be rewritten in a more compact way if the doer names are represented by a list (vector), or even better if they can be calculated having, for example, extensions as consecutive numbers (as in our case). For this latter case, the sequential program-summing contents of the variables N in the four doers may be as follows:

```
repeat(
  Fd += 1; Fd <= 4; direct # doer & Fd % '.'; Fsum += N
);
USER = Fsum
```

where frontal variable Fd keeps the current doer extension, which is attached to the word doer by a period using concatenation act &, and then vector-into-string merging act % with the given delimiter (here a period); this result being used as a current doer name in the hop between doers.

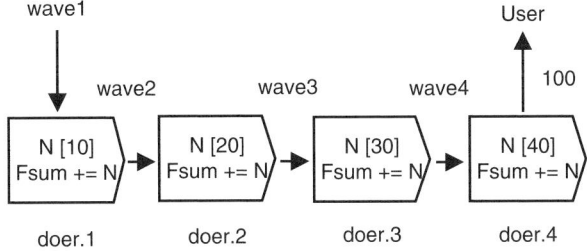

Figure 5.3. Sequential summation with migrating control.

The advantage of this program is that it can work sequentially for any number of doers whose names are organized in such a way. The output of the final result will be done from the last visited doer (here `doer.4`), from which the `repeat` construct fails as the received new value of Fd in it (i.e., 5) exceeds its given limit. Upon termination of the `repeat` construct, the frontal variable Fsum will keep the value that resulted from the last successful spatial iteration (which brought control into `doer.4`).

Using Central Doer for Summation. All the summing may also be done in another doer that keeps central control. This doer may request the four mentioned above doers holding vector elements and receive these elements from them for the summation, thus using the latter doers as a parallel and distributed memory only. Activation of the doers holding vector elements and bringing the elements to the central doer to be summed up may be done in any order or in parallel, by the following program:

```
sequence(
   (direct # any; Ftransit = N;
    BACK #; Nsum += Ftransit
   ),
   USER = Nsum
)
```

The work of this program is shown in Figure 5.4, where the first branch of `sequence` broadcasts from the central doer to the four doers in parallel.

The program then picks up the elements in N in each doer into a frontal variable Ftransit and brings it back to the central doer (using environmental variable BACK), to be added to the nodal variable Nsum in it. This branch first represents

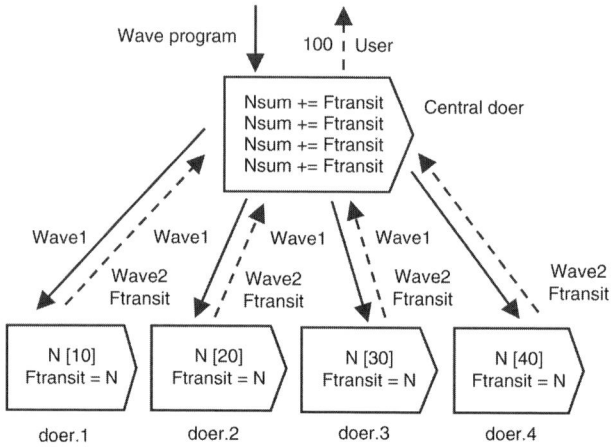

Figure 5.4. Vector summation with the use of a central doer.

Wave1 in Figure 5.4 and then reduces to Wave2, which will be as follows (accompanied in its movement by frontal variable Ftransit):

```
BACK #; Nsum += Ftransit
```

The second branch of sequence, working only after the full completion of the first branch, outputs the final value of Nsum from the central doer to the symbolic user.

USING RULE none. The overall control can be simplified, similar to the case considered before, by using echo rule none, leaving a single control point in the central doer after the summation process, as follows (using square brackets as the shortened representation of none):

```
[direct # any; Ftransit = N;
 BACK #; Nsum += Ftransit];
USER = Nsum
```

USING RULE grasp. Depending on the implementation, if the addition with assignment to Nsum is not considered as an indivisible operation, in order to preserve consistency if more than one such operation is performed in parallel on the same variable, the rule grasp should be used to suspend other operations in the node (here the central doer) until the current operation is completed, as follows:

```
[direct # any; Ftransit = N; BACK #;
  grasp(Nsum += Ftransit)];
USER = Nsum
```

CONTROLLED SEQUENTIAL SUMMATION. We can also perform explicitly controlled sequential summation in the central doer, as follows:

```
[sequence(direct # any; (BACK #; Nsum) += N)];
USER = Nsum
```

This program executes the broadcasting hop to all other doers sequentially, forming different branches spreading to the doers, which also include the replicated rest of the wave; and from these doers a direct adding of their personal element in N to Nsum in the central doer is performed, using environmental variable BACK to access Nsum remotely. As the summing in the central doer is performed sequentially, by turns for the branches, no protection of the variable Nsum from a simultaneous use by other branches (i.e., rule grasp) is required.

USING ECHO RULE sum. The special echo rule sum can also be used for this distributed computing example, where just naming variable N leaves its value open in each doer, to be picked and summed up by the internal distributed and parallel echo process in the most efficient way:

```
USER = sum(direct # any; N)
```

5.2 VIRTUAL WORLD PROGRAMMING

So far we wrote and executed wave programs in single or multiple doers, programming on the level of the execution world, treating doers as EW nodes—similar to what we usually do when using computers or computer networks. In this section, we will show how to work with a higher level abstraction, or virtual worlds (VW), which can be arbitrarily distributed between multiple doers and which can be treated (and also can exist) as an independent substance. We will then work on (and in) these VW directly, ignoring their distribution in EW (actually forgetting about the existence of the latter at all).

5.2.1 Creating Virtual World as a Knowledge Network

Let us create a simple VW example as a knowledge network as shown in Figure 5.5, with nodes a, b, c, and d and links p, q, r, s, and t (p, q, and s being oriented and r and t nonoriented).

Template-Based Distributed Network Creation. This may be done by the following program where rule create embraces a graph template based on a depth-first spanning tree covering the network of Figure 5.5, with node b as its root [a variety of techniques in WAVE for creating arbitrary networks in a distributed mode is considered in detail in our previous book (Sapaty, 1999a)]:

```
create(
    direct # b; + q # c;
    (+ s # a; - p ## b), (t # d; r ## b)
)
```

Act ## is used instead of # to express hops to the already existing node (here b), to which new links will be formed from the just created nodes a and d. A stepwise network creation process, where the contextlike rule create is inherited by parts of the wave after its gradual reduction and splitting, with a possible distribution between doers, is shown in Figure 5.6 (using allowed abbreviations of the language

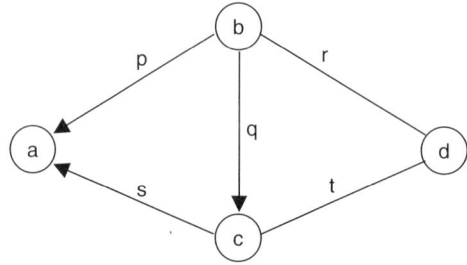

Figure 5.5. Virtual world as a knowledge network.

VIRTUAL WORLD PROGRAMMING

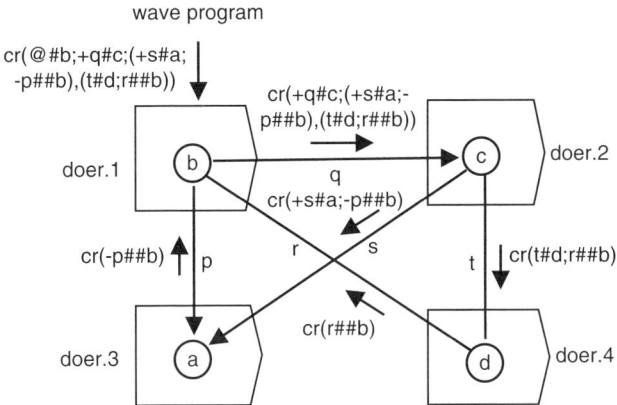

Figure 5.6. Stepwise parallel knowledge network creation and distribution.

constructs and also assuming doer.1 to doer.4 being default resources by the system).

A more economical solution for the creation of this network, which eliminates global search for the already existing node b, by first remembering its address in a frontal variable, say Fb, and then using it in the subsequent hops to b, may be as follows:

```
create(
  direct # b; Fb = ADDRESS; + q # c;
  (+ s # a; - p # Fb), (t # d; r # Fb)
)
```

Hopping to nodes by their addresses in WAVE-WP always indicates that these nodes must already exist (as addresses are formed only by the system during the creation of nodes), so the usual hop act # (instead of ##) may also be used in such cases, as in the example above.

Distribution Between Particular Doers. To provide a random distribution of the network only to the doers shown in Figure 5.6, their names may be given explicitly in the variable RESOURCES at the beginning of the program above, as follows:

```
RESOURCES = (doer.1, doer.2, doer.3, doer.4);
create(
  direct # b; Fb = ADDRESS; + q # c;
  (+ s # a; - p # Fb), (t # d; r # Fb)
)
```

To always provide the distribution of network nodes between the doers shown in Figure 5.6, the names of doers may be aggregated in the program with the node names, and the network creation program should be modified as follows:

```
create(
  direct # b_doer.1; + q # c_doer.2;
  (+ s # a_doer.3; - p ## b), (t # d_doer.4; r ## b)
)
```

5.2.2 Inhabiting the Virtual World with Mobile Entities

Having created the "tissue" of the virtual world as a distributed knowledge network, or KN, which can hold any information and in any form, a structural one included, we can animate this world by inhabiting it with some active creatures, or *entities*.

Simplest Case of Live Entities. These entities, in the simplest case, can just move between nodes through KN links, and stay in nodes for some time, say, 60 sec. The entities may randomly choose the next link to pass, in order to reach one of the neighboring nodes. The following program will do this by placing initially into nodes a, c, and d the same copy of the network navigation repetitive process (see Fig. 5.7):

```
direct # (a, c, d);
repeat(random(any # any); TIME += 60)
```

Seeing Other Entities. The functionality of mobile entities can be easily (and arbitrarily) extended, allowing them, for example, to see each other and inform the

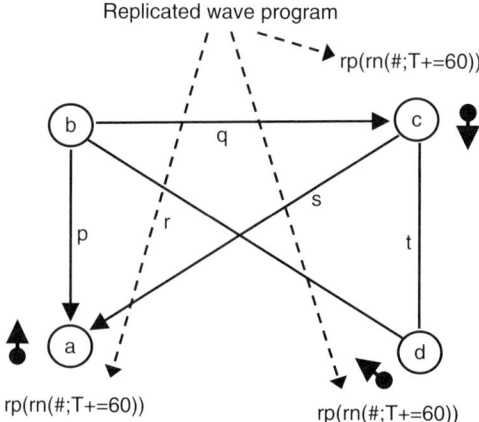

Figure 5.7. Inhabiting the virtual world with mobile entities.

user (or any other program or system) about this. And the entities may have personal names to be distinguishable from each other—as in the following program:

```
(direct # a; unit1),
(direct # c; unit2),
(direct # d; unit3); Fname = VALUE;
repeat(
  random(any # any);
  [Names != nil; USER = Fname & sees & Names % ''];
  Names &= Fname; TIME += 60; Names :: Fname = nil
)
```

The three entities are supplied with names `unit1`, `unit2`, and `unit3`, and each entity puts its name into vector `Names` in each node it visits and withdraws the name from `Names` when it leaves the node; and `Names`, as a nodal variable, is accessible by all entities currently staying in the node. Each entity entering a node outputs a message that it sees all entities listed in `Names` (if the latter is not empty). Rule `none` (its square brackets option) is used to simplify the program control. Entities stay in nodes for 60 sec and then randomly choose a neighboring node to go there, as before.

Communicating and Impacting Other Entities. It is easy to write in WAVE any scenarios where mobile entities not only see each other, as discussed above, but also can impact other entities, competing or cooperating. For example, we may allow entities to destroy other entities when they see them and let both destroyers and destroyed inform the user about these situations. This scenario for the three entities may look as follows:

```
(direct # a; unit1),
(direct # c; unit2),
(direct # d; unit3); Fname = VALUE;
repeat(
  random(any # any);
  [indivisible(
     Names != nil; N1 = Names : 1; Names : 1 = nil);
   USER = Fname & destroys & N1 % ''];
  Names & Fname; TIME += 60;
  or(
    (Names :: Fname == nil;
      USER = Fname & destroyed % ''; done !
    ),
    Names:: Fname = nil
  )
)
```

Having entered a node by random choice, an entity first checks nodal variable `Names` in it. If the latter is not empty, the entity removes the first element from this vector, informing a user that it destroys an entity with the removed name (i.e., only one entity can be destroyed by another entity at a node). Then the entity places its own name into Names in the current node and stays (sleeps) in the node for 60 sec, resuming work with checking its own presence in Names.

If the entity's name is missing from Names, it means that it has been removed from it by some other entity (i.e., the current entity has been destroyed), and the entity reports a user about this fact and terminates its work, otherwise continuing operation as described. In the program above, messages sent by entities go to the users linked with the current doers (i.e., the ones holding VW nodes in which the entities are currently staying). If needed, any central user can be easily organized to receive all messages from the whole distributed world (if this can be practical and not a bottleneck, of course).

Any other payload can be added to the mobile entities described above, with many such examples considered in Sapaty (1999a) so we will not elaborate this here.

5.2.3 Providing Openness of the Virtual World

We can make any changes to the described virtual world at run time, modifying or extending it, without termination or even suspension of its overall activity. The following program adds to the network of Figure 5.7 two new nodes: e (to be connected to node a by link g), and f (to be connected to nodes a and d, correspondingly, by links w and v). It also adds new link h between nodes a and d and deletes the existing link s between nodes a and c. The program also puts two new mobile entities into the new nodes e and f:

```
sequence(
  ((direct # a; (s # any; LINK = nil),
              create(g # e, h ## d)),
   (direct # d; create(v # f; w ## a))
  ),
  (direct # (e, f);
   repeat(random(any # any); TIME += 60)
  )
)
```

The resultant world, along with the steps of its modification, is depicted in Figure 5.8, where the new nodes e and f may be allocated to any new doers or to the same ones (i.e., doer.1 to doer.4).

The assignment of new nodes to particular doers may be also set up explicitly, as in one of the previous programs. The added entities can move now throughout the whole virtual network, along with the three previous ones shown in Figure 5.7.

The new entities can also be made more complex, say, by seeing and impacting each other, as described before. Any changes and extensions of the virtual world can be done simultaneously from its different points, from different doers, and by different users, without termination or suspension of the world's activity (or its parts), providing full openness of the distributed worlds organized in WAVE-WP.

VIRTUAL WORLD PROGRAMMING

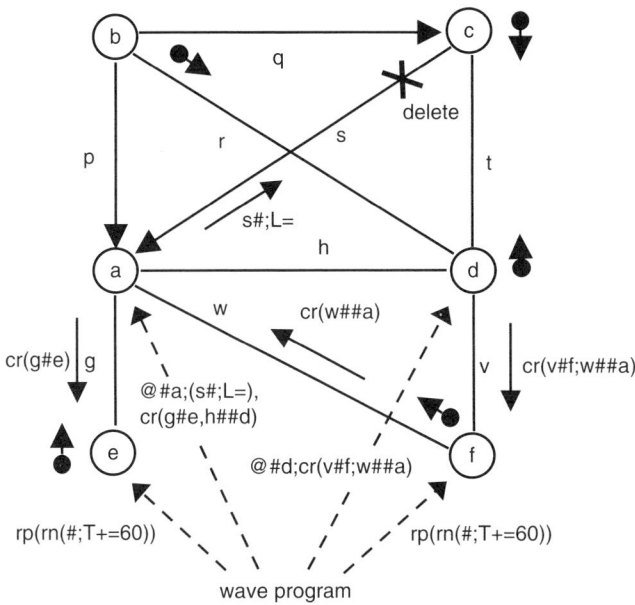

Figure 5.8. Run time modification and extension of the active virtual world.

5.2.4 Observation of the Virtual World

Despite the fully distributed nature of the virtual world described above, any of its observations in order to understand its behavior and assess and classify situations in it (which may be distributed too, covering many nodes) can be organized in WAVE-WP. This observation may be both local and global and from single or from multiple points. We will consider here only some simplest cases.

Direct Broadcasting to All Nodes. Listing all existing mobile entities throughout the whole system can be done by the following program, which first broadcasts from the start node (let it be a) to all VW nodes (including the start node itself) and then leaves the content of the variable Names in each of them open, to be collected and merged in one vector and then returned to and printed in the start node. (Names in each node keeps names of entities currently staying in the nodes, as before.)

```
direct # a; USER = (direct # all; Names)
```

If names of mobile entities need to be printed together with the names of nodes currently hosting them, the following program will do this (preventing from printing empty lists of entities):

```
direct # a;
USER = (direct # all; Names != nil; Names & CONTENT % '_')
```

All lists of entities in nodes, each followed by the corresponding node name, will be printed as a single vector of aggregate values (aggregation uses the underscore as a separator).

Spanning Tree-Based Collection. Instead of a broadcasting hop directly to all VW nodes for collecting names of possible entities in them, a more economical breadth-first spanning tree observation of the world can be organized. Starting from a single node (let it be b) such an observation gradually covers the whole world, with a subsequent collection of entity names in the nodes reached via the created spanning tree, as follows:

```
Fp = {
  grasp(N == nil; N = 1);
  (Names != nil; Names & CONTENT % '_'),
  (any # any; ^ Fp)
};
USER = (direct # b; ^ Fp)
```

The program is based on a recursive breadth-first navigation procedure Fp, which broadcasts by each of its invocations to neighboring nodes only, marking the passed nodes (using nodal variable N) to prevent their reentering by other branches. The rule grasp provides consistency of the marking operation, as more than one process may try to seize the node concurrently. This spanning tree observation of the world is depicted in Figure 5.9 (a possible breadth-first spanning tree of the network is shown boldfaced).

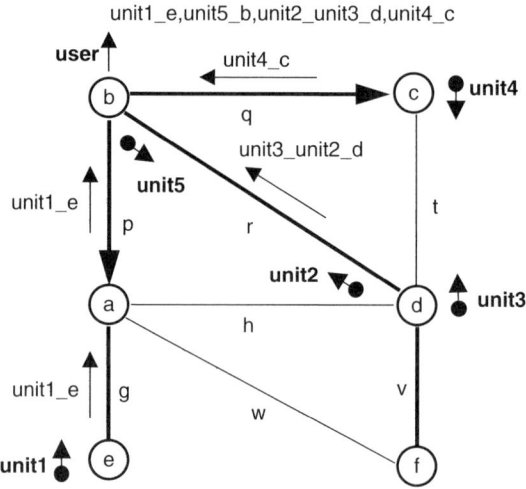

Figure 5.9. Spanning tree-based observation of the virtual world.

The following possible answer may be sent to the user:

```
unit1_e, unit5_b, unit2_unit3_d, unit4_c
```

Multipoint Observation. The observation of the world can also be organized in many points (say, a, b, and c) simultaneously, as by the following program (using the simplest case, where only the names of entities are collected):

```
direct # (a, b, c); USER = (direct # all; Names)
```

The output of names of all existing entities will be done independently in each of the three nodes (possibly, representing different users), where the same names may be collected in a different order from different starting nodes. The observation process put into each node can also be made continuous and regular (say, with an interval of 2 min), by the following modification of the program:

```
direct # (a, b, c);
repeat(USER = (direct # all; Names); TIME += 120)
```

Selected Observation. The situation may occur where from different nodes only some, selected, parts of the system may need to be regularly observed, rather than the whole system—say, nodes b and e from a, and nodes c and f from d, as follows:

```
(direct # a; N = b & e), (direct # d; N = c & f);
repeat(USER = (direct # N; Names); TIME += 120)
```

Examples of a more complex observation of distributed networked systems in WAVE-WP, including copying and return of the entire network topology in the language syntax, can be found in Sapaty (1999a), in relation to the predecessor language WAVE being a subset of the current language.

5.2.5 Distributed Inference in the Virtual World

Complex logical problems can be solved in distributed virtual worlds using the spatial pattern-matching nature of WAVE-WP. We will show here how a simple graph pattern, as a triangle, can be matched with the network shown in Figure 5.5, finding substitutions for variables X, Y, and Z in the pattern's nodes. The links of this pattern may have orientations and names or may be arbitrary as shown, respectively, in Figures 5.10a and 5.10b.

Finding substitutions for the variables in nodes may be done using a linear template of the pattern developing in the network stepwise, with direct coding of the template in WAVE-WP. The pattern leads to a solution (which can be more than one) only if all the three variables obtain values in the matching process, and in the order established [more about different pattern-matching techniques can be found in Sapaty (1999a)].

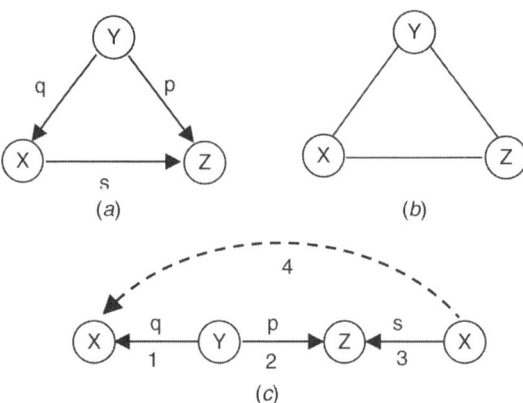

Figure 5.10. Simple graph pattern with free variables: (a) with named and oriented links, (b) with arbitrary links, and (c) a linear template of the pattern in (a).

Using a Pattern with Named and Oriented Links. The inference program in WAVE-WP for the pattern in Figure 5.10a with link names and orientations, based on the linear template shown in Figure 5.10c, may be as follows:

```
direct  # any; Fresult = CONTENT;
  - q   # any; Fresult &= CONTENT;
  + p   # noback; Fresult &= CONTENT;
  - s   # Fresult : 1; USER = 'XYZ:' & Fresult % ''
```

The program starts from all network nodes and then tries to grow the needed pattern on the network body in parallel, by stepwise (steps numbered in Fig. 5.10c) moving through the ordered links of the pattern, accumulating the match of variables X, Y, and Z in the frontal variable Fresult. The program ultimately prints the following result (from the node-matching variable X), which is unique for the pattern in Figure 5.10a and the network of Figure 5.5:

```
XYZ: c b a
```

Using a Pattern with Arbitrary Links. The wave program for the pattern given in Figure 5.10b, with any links between the nodes, will be as follows, providing a pure topology match of the triangle with the network of Figure 5.5 (assuming that the matching template also starts from X):

```
direct # all; Fresult = CONTENT;
  any  # any; Fresult &= CONTENT;
  any  # noback; Fresult &= CONTENT;
  any  # Fresult : 1; USER = 'XYZ:' & Fresult % ''
```

The program, initially applied to all nodes of the network in Figure 5.5 and being split at each stage simultaneously grows on the network body 12 possible solutions, outputting the following results for X, Y, and Z (the output for different matches may appear in different nodes—the ones in which the match is completed successfully (being also the starting nodes, i.e., corresponding to X):

```
XYZ: a b c    XYZ: a c b    XYZ: b a c    XYZ: b c a
XYZ: c a b    XYZ: c b a    XYZ: b c d    XYZ: b d c
XYZ: c b d    XYZ: c d b    XYZ: d b c    XYZ: d c b
```

With minimum modifications, the program above can output the united solution (consisting of the 12 matches) in a single node—the node where the program initially applies, which can also be one of the nodes of Figure 5.5, as follows:

```
USER = 'XYZ' & (
   direct # all; Fresult = CONTENT;
      any # any; Fresult &= CONTENT;
      any # noback; Fresult &= CONTENT;
      any # Fresult : 1; Fresult % ''
)
```

The output from this program will be in the form of a vector, for which the explaining sequence of variable names XYZ is exhibited only at the beginning of the resultant vector, as its first element and not prefixing each local solution as in the example before, as follows:

```
(XYZ, a b c, a c b, b a c, b c a, c a b, c b a,
      b c d, b d c, c b d, c d b, d b c, d c b
)
```

Any patterns for parallel inference by spatial matching in the virtual worlds can be represented in WAVE-WP [using standard templates directly expressible in the language syntax—see more in Sapaty (1999a) for WAVE], allowing us to solve complex logic problems in a fully distributed mode. Traditional logic programming languages, such as Prolog, for example, are easily convertible into this distributed and parallel pattern-matching mode.

5.2.6 Mobility of the Virtual World in the Execution World

The distribution of the created virtual worlds between doers may not be permanent and may change at run time. Different VW nodes, together with any local information associated with them and any processes currently staying in the nodes (such as mobile entities described above), can move freely between doers.

The moving nodes are preserving their links with other nodes that can be located in the same or in other doers and that can also be on the move between doers. For example, the following program applied in all nodes of the VW shown in Figures 7.8 and 7.9 provides an independent migration of VW nodes between the four doers mentioned above, where a next doer is chosen randomly from their list in the nodal variable Ndoers, and nodes are allowed to stay in doers for 2 min:

```
direct # any; Ndoers = (doer.1, doer.2, doer.3, doer.4);
repeat(TIME += 120; DOER = random(Ndoers))
```

All other activities in this VW, including movement of entities between nodes and their interaction and communication with users, as well as run time changes of the network topology, may take place simultaneously with the movement of the whole network or its parts between doers.

5.3 MOBILITY OF DOERS IN PHYSICAL WORLD

Doers, as already discussed in the book, may not only be networked computers but may also have physical bodies and freely move in physical space, communicating with each other. These mobile platforms may be of ground, underground, surface, underwater, air (say, fixed wing, rotary, rocket propelled, etc.), or space types. In WAVE-WP, we can manage any movement of doers in the physical world, both individual and group ones, where in the latter case doers may behave cooperatively or competitively.

5.3.1 Movement of a Single Doer

The following program, when put into any doer, organizes its continuous movement in PW with proper speed and with passing waypoints defined by a regular shift of coordinates x and y on given values (such as dx22 and dy11), using environmental variable WHERE to update the PW position:

```
SPEED = 67; repeat(WHERE += x22 y11)
```

This regular movement of a single doer is depicted in Figure 5.11.

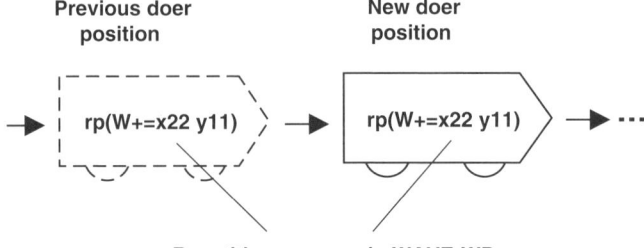

Figure 5.11. Regular movement of a single doer.

5.3.2 Free Movement of Multiple Doers

A variety of simultaneous movements in PW of multiple doers can be organized in WAVE-WP too. Elementary extension of the program above, by putting it initially into all available doers (let these again be doer.1 to doer.4, but now as mobile platforms), causes their simultaneous movement, as shown in Figure 5.12:

```
direct # any; SPEED = 67; repeat(WHERE += x22 y11)
```

The frontline in Figure 5.12 linking the doers may change shape in time, as the doers are moving without any external or mutual coordination, and conditions for movement of different doers in the real world may differ despite the same initial parameters set up in each of them.

5.3.3 Synchronized Movement of Multiple Doers

It is easy to establish a synchronized movement using a central doer, assumed to be separate from the four doers engaged in the movement, by the following modification of the program above to be loaded into this central doer:

```
repeat([direct # any; SPEED = 67; WHERE += x22 y11])
```

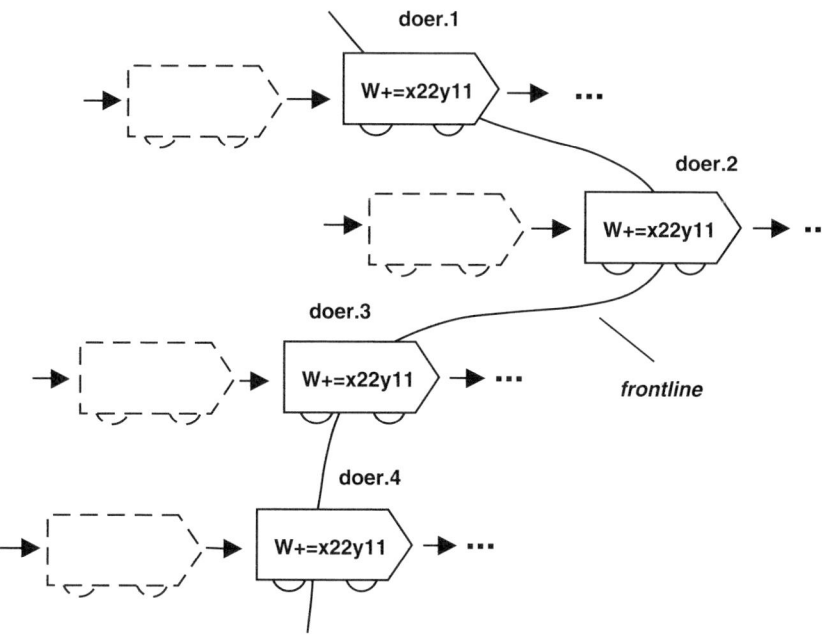

Figure 5.12. Simultaneous movement of multiple doers.

This program repeatedly activates parallel steps of movement of the four doers from the central doer in a broadcasting hop, where each new step starts only after the previous step is completed by all doers (this synchronization is guaranteed by the semantics of the rule `repeat`). So the current shape of frontline in Figure 5.12 may persist on average, with local deviations only during the waiting periods within steps.

With a minimal change, this program can also be used when one of the four doers engaged in the movement is also chosen to be the central doer, into which the program is loaded from the beginning:

```
repeat([direct # all; SPEED = 67; WHERE += x22_y11])
```

The central doer will manage the rules `repeat` and `none` (the latter represented by its square brackets option), regularly loading into every doer (including itself) the following shift-making wave:

```
SPEED = 67; WHERE += x22 y11
```

The code movement between doers can be reduced to a minimum if the wave above is recorded preliminary in each doer as a procedure, and then only activated from the central doer in each step of the movement, as follows:

```
[direct # all; Np = {SPEED = 67; WHERE += x22 y11}];
repeat([direct # all; ^ Np])
```

5.3.4 Movement of Multiple Doers by Turns

Any other forms of coordinated movement in PW may be written and implemented in WAVE-WP. As another simple example, let us consider the movement of doers by turns, one after the other (where the next doer can perform a current step only after the previous doer completes it), in a round robin fashion, with migrating control between doers starting from `doer.1`:

```
[direct # any, nil; Np = {SPEED = 67; WHERE += x22 y11}];
repeat(
  Fd += 1; or(Fd <= 4, Fd = 1);
  direct # doer & Fd % '.'; ^ Np
)
```

The sequential rule `or` has two branches, with the first branch just doing nothing if the growing doer extension in frontal variable `Fd` does not exceed the total number of doers (four in our case, being also the last doer's number), and the second branch returning `Fd` value to one otherwise, providing rotation of doer names within the distributed cycle. Such a movement is depicted in Figure 5.13, where the frontline of the positions of doers in space may persist on average, changing only within steps

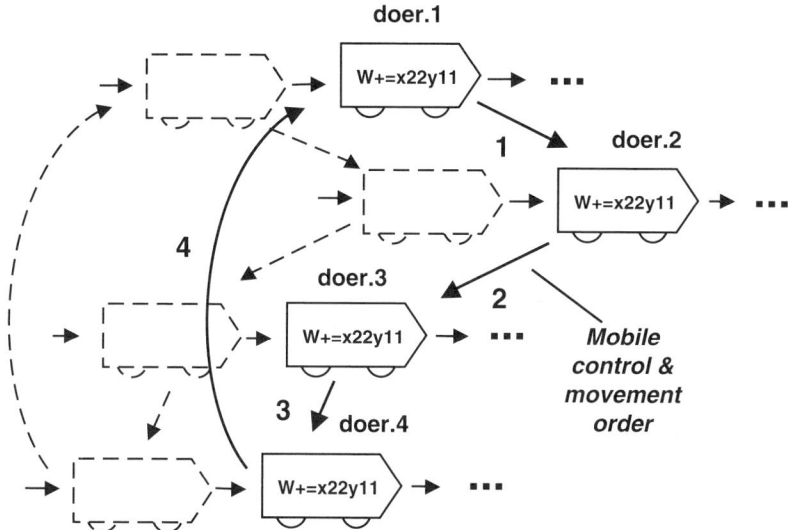

Figure 5.13. Movement of doers by turns.

of the distributed cycle covering sequentially all the four doers. This program may be initially loaded into any doer, which will immediately transfer control into doer.1 actually starting the movement.

We can modify the program in a way that it will allow any doer, into which the program is loaded initially, to start the movement itself. For this, the program should find the current doer name extension first (using environmental variable DOER), to be used as the initial value of variable Fd, not zero as before (on default), as follows:

```
[direct # all; Np = {SPEED = 67; WHERE += x22 y11}];
Fd = DOER | '.': 2;
repeat(
    ^ Np; Fd += 1; or(Fd <= 4, Fd = 1);
    direct # doer & Fd % '.';
)
```

5.3.5 Adding Payloads to Mobile Doers

Any payload may be added to the movement of doers, for example, letting them dig a hole after each step (say, with a diameter of 0.5 m and a depth of 1 m), and pick up the soil from the hole into a container attached to the doer, to be released at the end of movement at the distance of an additional step. Such a scenario for a single doer (allowed to make 10 holes) may look as follows (it may be easily extended for multiple doers, similar to the examples above):

```
SPEED = 67;
repeat(
  Nsteps < 10;
  Ncontainer += (0.5, 1) ? digPickup;
  Nsteps += 1; WHERE += x22 y11
);
Ncontainer ? release
```

In this program, an external function `digPickup` (assumed to be the hardware part of the doer) is used. It returns physical matter as a result (here soil), being added to the content of the variable `Ncontainer` holding matter rather than information, which, by another external function `release`, is emptied after the allowed number of steps (here 11, the last one without the digging) expires.

5.4 MOVING AND ACTING IN PHYSICAL WORLD DIRECTLY

So far we have been moving in the physical world and making operations in it (even thinking about the movements and operations) *from the position of sitting in a doer and via a doer*. This reflects the traditional philosophy and culture, where to do or organize something in PW, robots are explicitly programmed for certain operations and communication with other robots, so *we are dealing with PW indirectly*.

But WAVE-WP also allows us to work on a much higher level of abstraction and concentrate on semantics of the movement and operations to be performed in PW (on both physical matter and information), rather than on programming robots that would perform the movement and operations. This may provide a good deal of freedom to how these operations can be performed by mobile hardware, and how many mobile units should be engaged, where the mapping of operations onto the doers may be dynamic and depending on their current availability.

An overall control over the operations in PW may also be on a pure semantic level, without details of implementation, just hinting on its general (say, sequential or parallel) nature. Let us consider some examples of direct programming in PW, just assuming that we have an unlimited power to stay in any of its point (or points, simultaneously), move to any other points, do any jobs, and get anything we want.

Imagine that we have a set of points in PW to be visited, as follows, where in each point a certain PW parameter should be checked:

```
Fpoints = (x21 y41, x21 y42, x21 y43,
           x22 y41, x22 y42, x22 y43,
           x23 y41, x23 y42, x23 y43)
```

The result of this search should be the maximum value of this parameter among all these points, together with the coordinates of the point giving the maximum

MOVING AND ACTING IN PHYSICAL WORLD DIRECTLY

(or one of them, say, visited first, if the maximum repeats). Different search strategies to perform this task can be expressed in WAVE-WP.

5.4.1 Sequential Movement in Physical World

Using Migrating Control. Using control migrating directly in PW between the points, with removal of used elements from frontal variable Fpoints holding a vector of coordinates of the points, the solution may look as follows:

```
Fpoints = (x21 y41, x21 y42, x21 y43,
           x22 y41, x22 y42, x22 y43,
           x23 y41, x23 y42, x23 y43);
repeat(
  Fpoints != nil; direct # Fpoints : 1;
  Fpoints : 1 = nil; N = QUALITIES : 23;
  or((Fresult : 1 < N; Fresult = N & WHERE), nil)
);
USER = Fresult
```

The removal of the used elements in Fpoints is to simplify the solution here, but this also reduces the amount of information to be propagated with the control. The program, starting in any point of physical or virtual world, visits the points in the given order (by WAVE-WP semantics, temporary nodes are being created in the points visited), disappearing when control moves to other points (nodes).

The parameter of interest is assumed to be the 23rd element of the vector of PW parameters accessible in each point by environmental variable QUALITIES. The frontal variable Fresult always holds the best so far (maximum in our case) parameter value, as its first element, with the second element being the coordinates of the point that defined this value first. If the visited node gives a better parameter value, the whole content of Fresults changes to reflect this new point. After visiting the last point in Fpoints (with the latter becoming empty, or nil, eventually), the result accumulated in Fresult is sent to the user, as shown in Figure 5.14.

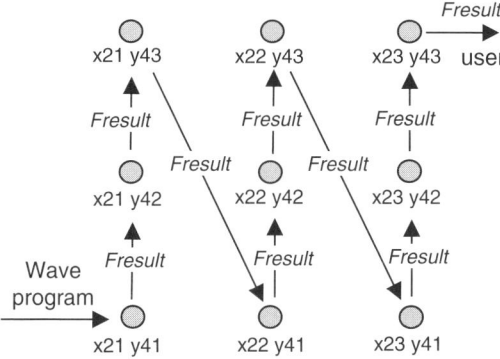

Figure 5.14. Sequential movement in physical world with migrating control.

Using Centralized Control. We can rewrite the program above so that the overall control does not migrate between the PW points but rather stays in some central point, which may be any imaginable PW point or VW node, where the program is initially applied, with the given points visited sequentially, in the given order, each time starting from this point. This will always return the information from any point to the central point for its checking and updating of the result, accumulated in the center. The program solution may be as follows:

```
Npoints = (x21 y41, x21 y42, x21 y43,
           x22 y41, x22 y42, x22 y43,
           x23 y41, x23 y42, x23 y43);
sequence(
   sequence(
     direct # Npoints;
     Ftransit = QUALITIES : 23 & WHERE;
     BACK #; Ftransit : 1 > Nresult : 1;
     Nresult = Ftransit
   ),
   USER = Nresult
)
```

As the visiting of all points is from the central point, nodal variable Npoints is used in it instead of the frontal Fpoints, as there is no need now to move the information between nodes.

The broadcasting hop direct # Npoints from the central point will be split and sequentialized by the second (inner) branching rule sequence, where the hop to a new point will be allowed only after the full completion of all processing from the previous point. Frontal variable Ftransit is used to bring back to the central point the parameter value (together with the coordinate of the PW point), assisted by environmental variable BACK. The general organization for this solution is shown in Figure 5.15.

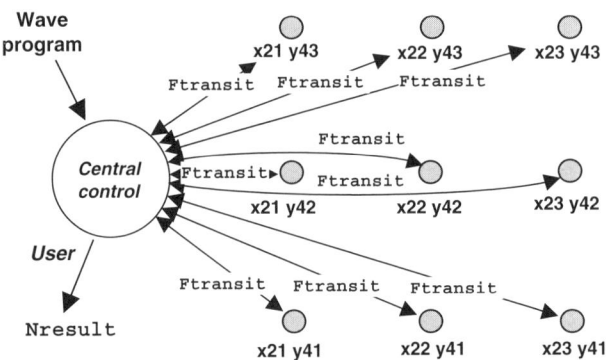

Figure 5.15. Investigation of physical world via the central control.

5.4.2 Parallel Movement in Physical World

With some changes, the previous program can also visit all the PW points in parallel, accumulating the solution in the central point concurrently. For this, the inner `sequence` rule should be removed, and operations in the central node, after bringing back the result from each PW point, should be protected by rule `grasp`, as they may be attempted to be executed simultaneously with similar operations coming from other points. The resultant program follows:

```
Npoints = (x21 y41, x21 y42, x21 y43,
           x22 y41, x22 y42, x22 y43,
           x23 y41, x23 y42, x23 y43);
sequence(
  (direct # Npoints;
    Ftransit = QUALITIES : 23 & WHERE; BACK #;
    grasp(
      Ftransit : 1 > Nresult : 1;
      Nresult = Ftransit
    )
  ),
  USER = Nresult
)
```

The general picture of the work of this program in a distributed space will resemble the previous one, depicted in Figure 5.15.

Representing a combination of the needed parameter value and physical coordinates of the nodes as an aggregate value, we can also directly use echo rule `max`, bringing the simplicity of this program to its limit:

```
Npoints = (x21 y41, x21 y42, x21 y43,
           x22 y41, x22 y42, x22 y43,
           x23 y41, x23 y42, x23 y43);
USER = max(direct # Npoints; QUALITIES : 23_WHERE)
```

The rule `max` will assess the aggregate values by their first component only (here the parameter value), while the rest of the aggregate (i.e., the node coordinates) will be considered just as an attachment to the first component, due to the semantics of this rule.

5.4.3 Combined Sequential–Parallel Movement

We have examined some extreme cases of visiting the given points in PW: sequential with distributed migrating control and both sequential and parallel with centralized control. Any solutions between these extremes may be of interest too—as a rational structuring and comprehension of activities in the real world.

We will consider the case where sequential and parallel navigation and investigation of the world go side by side; for example, where the search on x coordinate is parallel and centralized and on y coordinate is sequential and migrating. (Of course, this is a very artificial case but is used here to show the programming techniques in WAVE-WP only, hinting on how to apply it for more complex cases.)

We shall also write a program for any number of points to be visited, assuming for clarity that moving on both coordinates is by regular intervals, where Fx, FX, and Fdx are frontal variables keeping, correspondingly, start, final, and step values on the x coordinate (similarly, Fy, FY, and Fdy for coordinate y).

The following program performs movement in PW sequentially on coordinate y, using procedure Fpry, and in parallel on coordinate x, using procedure Fprx. The latter involves recursively as many copies of itself and Fpry (the latter sequentially covering all values of y) as there are steps on x, as follows:

```
Fx = x20; FX = x23; Fdx = 1;
Fy = y40; FY = x43; Fdy = 1;
Fpry = {
  repeat(
    Fy <= FY; direct # Fx Fy; N = QUALITIES : 23;
    or((Fresult : 1 < N; Fresult = N & WHERE), nil);
    Fy += Fdy
  );
  direct # Fstart;
  grasp(Fresult : 1 > Nresult : 1; Nresult = Fresult)
};
Fprx = {
  (Fx, Fy, FY, Fdy) ^ Fpry,
  (Fx += Fdx; Fx <= FX; any ^ Fprx)
};
Fstart = ADDRESS;
sequence(any ^ Fprx, USER = Nresult)
```

The number of steps on both coordinates x and y are not known in advance, and these steps are involved dynamically (recursively for x and repetitively for y), until the coordinate limits given by FX and FY are reached.

The address of the central point, to which partial results obtained on the whole coordinate y for different values of x should be returned, is kept in frontal variable Fstart. Upon termination of Fpry, after reaching the limit on coordinate y for a certain value of x, the local result accumulated in variable Fresult is sent back to the central point to be compared there with the best so far result accumulated in nodal variable Nresult, updating the latter if needed.

Procedure Fpry inherits frontal variables Fx and Fy with starting coordinate values (Fx being different for each invocation of Fpry), final value of y in FY, and the step Fdy. Procedure Fprx invoking both Fpry and itself inherits all frontal variables available prior its invocation, this being indicated by any as its left operand. After the execution of all parallel invocations of the migrating control on coordinate y for different x, the program outputs the final result to the user, as shown in Figure 5.16.

MOVING AND ACTING IN PHYSICAL WORLD DIRECTLY

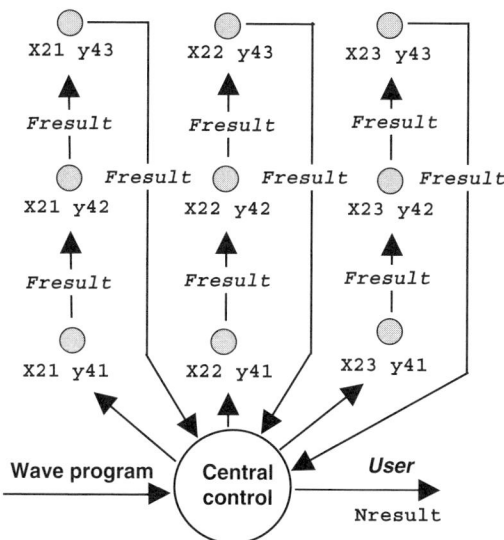

Figure 5.16. Combined sequential and parallel navigation.

Use of Echo Rule max. Using echo rule max, and also the fact that indexing act : can be applied to both vectors (for accessing their elements) and aggregate values (for addressing their constituent parts), the program above can be effectively simplified, as follows:

```
Fx = x20; FX = x23; Fdx = 1;
Fy = y40; FY = x43; Fdy = 1;
Fpry = {
  repeat(
    Fy <= FY; direct # Fx Fy; N = QUALITIES : 23;
    or((Fresult : 1 < N; Fresult = N_WHERE), nil);
    Fy += Fdy
  ); Fresult
};
Fprx = {
  (Fx, Fy, FY, Fdy) ^ Fpry,
  (Fx += Fdx; Fx <= FX; any ^ Fprx)
};
USER = max(any ^ Fprx)
```

After termination of each sequential process, the resultant value in Fresult is left open in the final PW points reached, and all these values will be echoed and merged in parallel by the rule max embracing the whole parallel and distributed, spatial process.

5.4.4 Adding Payload: Planting Trees

The sequential, parallel, and sequential–parallel control structures described above, allowing us to navigate in PW directly, may have different payloads, more complex than just finding the maximum of a PW parameter considered before. We will show an example of how the latest one, sequential–parallel control can be used for a scenario of planting trees in the visited points of PW.

Before starting the sequential processes along the coordinate y, in parallel for all cases of x, the number of future hops along y (to be the same for each sequential process) is calculated, and the corresponding number of trees is picked up in the frontal variable Ftrees. This number of trees will then move in space with one tree planted (and therefore removed from the stock) in each point visited. The program will be as follows:

```
Fx = x20; FX = x23; Fdx = 1;
Fy = y40; FY = y43; Fdy = 1;
Number = ((FY - Fy / Fdy) ? integer) + 1;
Fpry = {
  Ftrees = "tree" * Number;
  repeat(
    Fy <= FY; direct # Fx && Fy;
    (Ftrees, 1) ? plant; Fy += Fdy
  )
};
Fprx = {
  (Fx, Fy, FY, Fdy) ^ Fpry,
  (Fx += Fdx; Fx <= FX; any ^ Fprx)
};
sequence(any ^ Fprx, USER = planting_completed)
```

The needed number of trees is picked up into the frontal variable Ftrees (holding physical objects rather than information), and the external procedures plant plants one tree from Ftrees in each PW point visited, automatically reducing the number of trees in Ftrees.

We may rewrite this program without the use of procedures Fprx and Fpry, as one integral spatial module, just to show some other possibilities of the language. For this, all x coordinates of the PW points to be visited are calculated in advance and stored in a nodal variable Nxx and then used in parallel by simultaneous access to all elements of the vector Nxx allowed by the language (by just naming the vector as a move). The modified program will look as follows:

```
Fx = x20; FX = x23; Fdx = 1;
Fy = x40; FY = x43; Fdy = 1;
repeat(Fx < FX; Nxx &= Fx; Fx += Fdx);
Number = ((FY - Fy / Fdy) ? integer) + 1;
[Nxx; Fx = VALUE; Ftrees = "tree" * Number;
 repeat(
```

```
      direct # Fx && Fy;
      (Ftrees, 1) ? plant; Fy += Fdy
    )];
USER = planting_completed
```

As another program simplification, the termination of each sequential process is supposed to be automatically provided by the external procedure `plant` itself. This gives termination state `fail` if there are no more trees to plant in the local mobile stock kept in `Ftrees`, thus terminating the loop embraced by the rule `repeat`. This program simplification will, however, cost one additional (and useless) move in physical space and may not be acceptable from efficiency reasons.

5.5 PROGRAMMING IN INTEGRATION OF PHYSICAL AND VIRTUAL WORLDS

In the previous sections, we considered examples of organization and processing of persistent discrete and networked virtual worlds, which could accumulate and hold arbitrary knowledge in a distributed form, and also continuous physical worlds with temporary nodes appearing only during the movement and operations in them.

Using WAVE-WP, as already discussed, we can combine the two worlds within one model and find efficient solutions of complex problems of different natures by the help of this integration. We will consider here only a very simple example of how the virtual world can keep distributed persistent knowledge representing at the same time the physical world, and how such a symbiosis can guide movement and operations in PW with the accumulation of further knowledge in VW.

5.5.1 Planting Trees in the United World

Taking as a basis the knowledge network in Figure 5.5, we will bind its nodes, during their creation, with certain PW locations (thus receiving the VPW, or PVW, kind of nodes), and substitute link names in Figure 5.5 with numerical weights. We will also add to this network four pure VW nodes (all having initially a zero content), connected to the VPW nodes by links named `planted`, for accumulation of some persistent information, as shown in Figure 5.17 (with VPW nodes shaded).

We will use a modified tree-planting scenario to demonstrate a possible use of the combined network of Figure 5.17. This scenario may look very artificial too and is used here only to show some spatial dynamics supported by the language (other payloads could make much more sense). We will inject into the network a number of independent mobile processes, which will be randomly roaming via the network links and planting trees in each PW point visited (these points represented by the VPW nodes).

Each such process will pick up and carry with it a certain number of trees, planting in each PW point visited their number corresponding to the weight (content) of the link passed, until this is possible (the number of trees to be currently planted cannot exceed their remaining mobile stock), terminating otherwise.

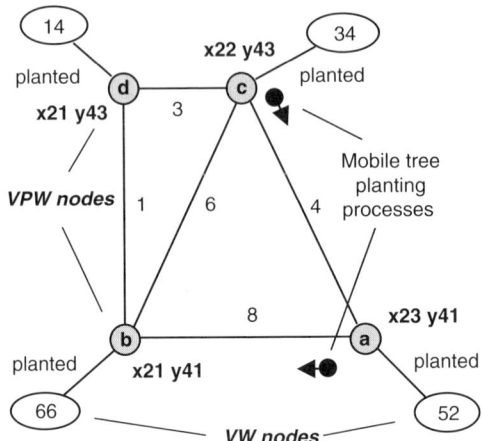

Figure 5.17. Integration of physical and virtual worlds in one model.

Having planted the needed number of trees in each VPW node, this number (i.e., the passed link weight) will be added to the content of the VW node reachable by link planted from the current node. These VW nodes thus serve as permanent counters of the planted trees, which may be accessed and shared by many mobile processes, differing from nodal variables, which are temporary and removed when the wave processes that created them terminate (N type) or after a system threshold cleaning period (M type).

A corresponding wave program, creating the network in Figure 5.17 and starting the mobile tree-planting processes simultaneously in nodes a and c, subsequently randomly moving between the VPW nodes with a certain speed (66 units), may be as follows. (It assumes that the initial stock of 100 trees for each self-replicating mobile process is available at the starting node, which may be outside the world shown in Fig. 5.17.)

```
[create(
  direct # b_x21 y41; 6 # c_x22 y43;
  (4 # a_x23 y41; 8 ## b), (3 # d_x21 y43; 1 ## b)
)];
[direct # any; create(planted # 0)];
Ftrees = "tree" * 100; direct # (a, c);
repeat(
  random(any_s66 # vpw); (Ftrees, LINK) ? plant;
  (planted # any) += LINK;
)
```

The created network of Figure 5.17 keeps the knowledge about PW locations and the allowed paths between them, as well as the number of trees to be planted after each hop, and the number already planted in different PW locations, in a convenient and fully distributed form. This persistent knowledge can be easily shared by different processes navigating the network.

5.5.2 Observation in the United World

Any observation of this united active world can be organized, telling us, for example, the total number of trees planted so far in a certain PW point by different migrating processes. The program for such an inquiry, say, in node d, may be as follows (the execution of the inquiry may work simultaneously with the tree-planting processes, not harming the latter):

```
USER = (direct # d; planted # any)
```

This will give the result 14 in the current snapshot shown in Figure 5.17.

The inquiry may be for more than one PW point, or for all of them (formally represented as VPW nodes) as follows, where also the names and coordinates of the corresponding VPW nodes will be printed together with (and before) the total number of trees planted in them:

```
USER = (direct # vpw; CONTENT_WHERE_(planted # any))
```

The printed result, corresponding to the snapshot shown in Figure 5.17, will be as follows:

```
a_x23 y41_52, b_x21 y41_66, c_x22 y43_34, d_x21 y43_14
```

Region Search. The observation may concern not only particular nodes but also a particular region in PW where the nodes may be located. Such a region may be defined by coordinates of its center and the search radius (say, x21.5, y42.5, and r1) as in the following request:

```
USER =
  (direct ## x21.5 y42.5 r1; CONTENT_WHERE_(planted # any))
```

which will enter nodes d and c only, with the collected and printed result:

```
d_x21 y43_14, c_x22 y43_34
```

Summary on a Region. We may, for example, get interested not in naming every node in the region with the local number of trees, but rather in some general information, such as, for example, the total number of trees planted in this region so far, as follows:

```
USER = sum(direct # x21.5 y42.5 r1; planted # any)
```

which will result in 48 for the snapshot depicted in Figure 5.17.

The united VPW world may be fully open and may be arbitrarily changed at run time, with many users and processes entering or leaving it simultaneously.

5.5.3 Programming of Spatial Dynamics

The network in Figure 5.17 has been considered stationary so far, with mobile processes roaming throughout it, where VPW nodes were always associated with the same locations in PW. But these nodes can be made mobile themselves while preserving links with each other and with the pure VW nodes accumulating the number of planted trees (the latter nodes, being PW free, are beyond any notion of physical movement).

For example, the following program, broadcasting to all VPW nodes and setting there a repetitive process changing a node's position in PW (on half a unit for both x and y), can make the whole distributed world in Figure 5.17 move in PW regularly (every 3 min), while preserving the internal logical structure of the united world and the life of the mobile tree-planting processes in it:

```
direct # vpw; repeat(JOINT += x0.5 y0.5 t180)
```

The environmental variable JOINT is used here to change the position of nodes in the united world on a number of parameters simultaneously (here location and time, whereas earlier we were using separately WHERE to change the place, and TIME to update the time).

The mobile processes may continue moving and planting trees in the VPW nodes, the latter are now moving themselves and changing positions in PW (this means that the trees will be planted in new physical world locations too). The VW nodes connected to VPW nodes by links planted will be reflecting the total number of trees planted *via* these VPW nodes, rather than the existence of trees in certain PW locations, as before. A snapshot of our united world moving in PW by the distributed program above is shown in Figure 5.18.

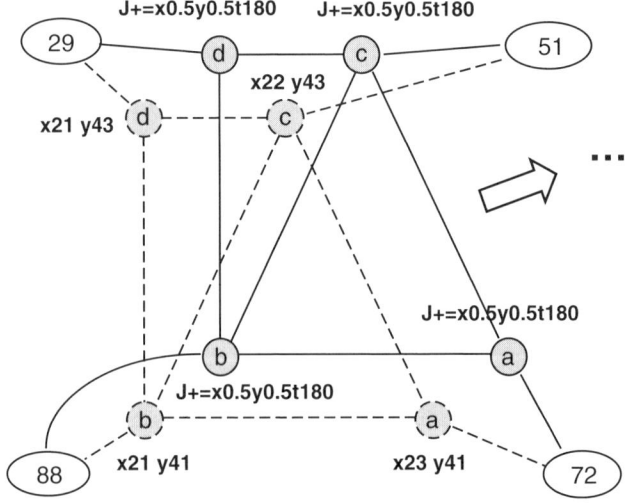

Figure 5.18. Mobility of nodes in the integrated world.

PROGRAMMING IN INTEGRATION

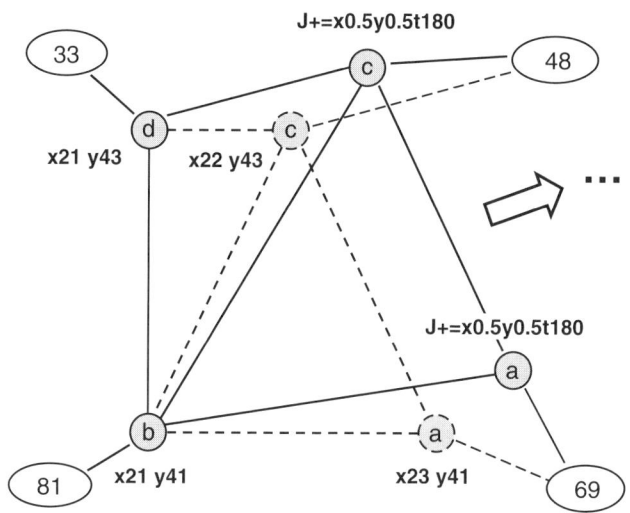

Figure 5.19. Coexistence of stationary and mobile nodes in the integrated world.

In a more general case, some nodes may be made mobile while others left stationary, as the following program does, supplying with the mobility nodes a and c only (see Fig. 5.19 for a snapshot):

```
direct # (a, c); repeat(JOINT += x0.5 y0.5 t180)
```

In addition to this full or partial dynamics of our united active world in PW, we may organize a run time movement of any or all its nodes not only in the physical world but in the execution world as well, between doers. The following program will do this regularly (randomly choosing a next doer to move into, from their list of eight, and with intervals of 4 min) for all nodes independently, including the four VW ones, regardless of their stationary or mobile nature in PW:

```
direct # any;
repeat(
   TIME += 240;
   DOER = random(doer.1, doer.2, doer.3, doer.4,
                 doer.5, doer.6, doer.7, doer.8)
)
```

This program leaves regular processes in each node, which will be working simultaneously with other stationary or mobile processes in the same and other nodes, which we discussed above, providing an absolute openness and freedom of evolution and migration of the united worlds organized in WAVE-WP.

5.6 CONCLUSIONS

We have investigated the use of the WAVE-WP model and language for programming practical scenarios from different worlds and on different system levels. We started from traditional calculations in single computers and their networks (i.e., from the execution world), then created active dynamic virtual worlds and solved simulation and pattern-matching problems in these worlds directly (regardless of their distribution between computers or doers).

We programmed movement and interaction of doers in the physical world (close to the traditional level of mobile robots) and then showed how to describe distributed problems in the continuous physical worlds on a semantic level, ignoring the execution world and providing maximum freedom for implementation. We have also shown on programming examples of how physical and virtual world may be deeply integrated within one model, with dynamic correspondence between the two worlds and the mobility of integrated model in both physical and execution worlds.

The high dynamics, flexibility, openness, and total freedom of movement, together with the possibility of merging different worlds and abstractions processed by the universal spatial WAVE-WP automaton, allow us to receive efficient solutions for traditionally heterogeneous and multilingual problems seamlessly, within the same simple formalism. WAVE-WP provides a unique opportunity to program the needed whole behavior of complex distributed systems not as a result of interaction of multiple components, which may often be unpredictable, but from the very beginning, as an integral, goal-driven spatial scenario dynamically matching the system to be controlled.

WAVE represents a sort of agent-free parallel and distributed programming, where stationary or mobile agents (software or hardware ones), or their equivalents, appear on the implementation level only, when required or available. This allows us to have clear and compact program solutions of a variety of complex simulation and management problems, as most of traditional space- and time-consuming routines are hidden within the language interpreter, relieving the application programming dramatically.

6

EXEMPLARY MISSION SCENARIOS

An expression in WAVE-WP of a number of exemplary mission scenarios in distributed physical world will be considered. Different forms of movement of hierarchically organized groups will be presented, where parallel movement by "subordinates first" may be used in combat operations and movement by "heads first" on a march to new positions through distributed terrain. Synchronized column movement may be useful for transportation via limited spaces. Integration of different forms of movement within the same scenario will be demonstrated.

Cooperative movement in physical space with certain physical payload and its remote and distributed processing, such as applying fire-extinguishing chemicals to a forest fire, will be exhibited. A variety of the description levels will be considered, ranging from top semantic to the detailed movement, interaction, and command and control of a robotic team moving in a column, with synchronized firing.

Distributed space search and processing, on an example of collective cleaning, guided by the distributed dynamic virtual world, will be described. Optimized solutions will be found in the virtual world first in a look-ahead simulation mode and subsequently used for guiding parallel movement and operations in the physical world. A variety of coordinated space-cleaning solutions, with local and global optimization, will be discussed.

Ruling Distributed Dynamic Worlds, by Peter S. Sapaty
ISBN 0-471-65575-9 Copyright © 2005 John Wiley & Sons, Inc.

Possible distributed operations on other planets for search and collection of soil samples by multiple rovers will be demonstrated. The virtual world will be used to create and process a distributed weighted map of the territory to be searched, assisting in dynamic finding of optimum routes for transportation of physical matter to the collection point.

6.1 COORDINATED MOVEMENT OF A GROUP

6.1.1 Stepwise Movement of a Two-Level Hierarchy

First, we will consider, on a pure physical world level, a coordinated movement of a hierarchically organized group of entities formed from two levels having a leader, or *head*, and a number of ordinary *members*. The entities may be of any nature (say, from dismounted soldiers or pieces of manned or unmanned mobile hardware to complex multiple-component organizations, e.g., armies).

Let us assume that by a command from the head of the group, its members perform certain movement in space, after which the head moves into a new position itself, then orders the members to move further again, and so on. We will consider the simplest case where the whole group, including the leader, moves within each step the same distance, in the same direction, and with the same speed. The wave program for this will be as follows:

```
Nhead = x25.5 y40.2; Fshift = dx10.0 dy0;
Frange = r1.5; Fspeed = s35;
direct # Nhead;
Nmembers =
  (x35.4 y25.2, x35.4 y35.2, x35.4 y45.2, x35.4 y55.2);
Nmembers =
  (direct # Nmembers; release(repeat(TIME += 60)); WHERE);
repeat(
  Nmembers =
    (direct ## Nmembers_Frange;
     SPEED = Fspeed; WHERE += Fshift; WHERE);
  TIME += 60; SPEED = Fspeed; WHERE += Fshift
)
```

The head and the four group members are put into their initial PW positions by coordinates given in variables, respectively, `Nhead` and `Nmembers`. This step also supplies all group members with a continuous process (just awakening every 60 sec) allowing these (usually temporary) p-type nodes to exist indefinitely.

The head of the group then collects actual coordinates of group members (which may differ from that initially planned in `Nmembers`—due to the limited precision of movement in PW), putting them again into variable `Nmembers`, and begins repetitive management of the group movement in space.

COORDINATED MOVEMENT OF A GROUP

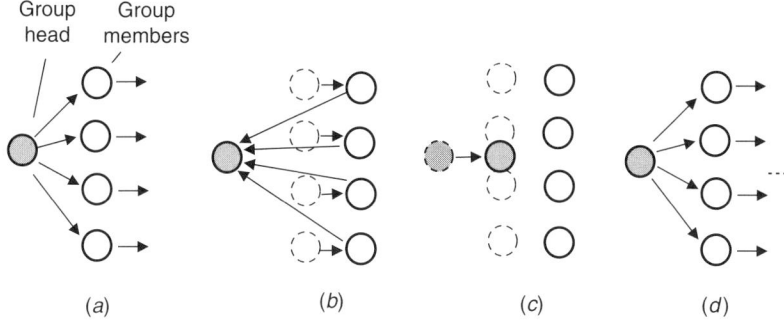

Figure 6.1. Coordinated movement of the two-level hierarchy: (a) issuing a command by the head to group members using their coordinates, (b) moving group members and collecting their new coordinates by the head, (c) moving the head, and (d) same as in (a).

During this movement, the head orders all group members to move from their current positions (on the same coordinate shift, given in the replicated frontal variable Fshift, and also with recommended speed in Fspeed), using fresh coordinates in Nmembers to contact them (along with the precision range held in frontal variable Frange). After completing the current step by all group members and collecting their new physical coordinates in Nmembers, the head, after time delay, makes its own move on the same shift and with the same speed, and so on, as shown in Figure 6.1.

This stepwise movement of the two-level hierarchy, where lower levels execute orders from higher levels and report the latter on their new locations in space, can be extended to any number of levels.

6.1.2 Creation of a Persistent Infrastructure and Moving with It

The solution above describes group nodes and their coordinated movement on a pure PW level, where PW nodes can be accessed by their physical coordinates only, and they must constantly keep active processes in order to exist (according to the semantics of p-type nodes of the language).

The group organization and command and control in and over it can be essentially simplified if to set up a persistent hierarchical infrastructure of the group (with persistent pv types of nodes and links between them). Such an infrastructure for two levels is shown in Figure 6.2.

Creation of this infrastructure in WAVE-WP with combined physical–virtual nodes (using the same initial node coordinates as before) will be as follows:

```
create(
  direct # 'head'_x25.5 y40.2;
  +'link' # ('member1'_x35.4 y25.2, 'member2'_x35.4 y35.2,
             'member3'_x35.4 y45.2, 'member4'_x35.4 y55.2)
)
```

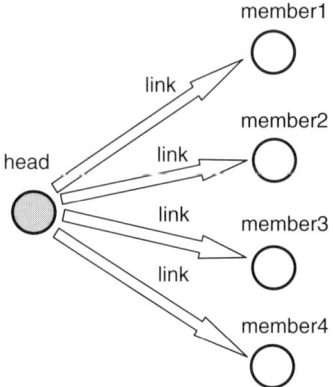

Figure 6.2. Two-level infrastructure set up upon the group.

All nodes are of pv type now, and the network is persistent regardless of the presence or absence of activities in its nodes. It always keeps updated coordinates of all nodes on the internal level, so there is no need to collect them explicitly each time when they change.

The repetitive movement of the whole group using this infrastructure can be expressed as follows:

```
Fshift = dx10.0 dy0; Fspeed = s35;
direct # 'head';
repeat(
  [sequence(+'link' # any, nil;
          SPEED = Fspeed; WHERE = Fshift)])
```

In the program above, the rule sequence splits the wave it embraces into two branches executed sequentially (decomposing the head and replicating and attaching the common tail), which implement this first-group-then-leader moving scenario.

6.1.3 Extending to Any Number of Layers

The previous program for a two-layer organization can be easily extended to serve a hierarchy with any number of layers, as follows (it is assumed that such a hierarchy has on its top the node head, with any other node names possible, and all links named link are directed from higher to lower layer nodes):

```
Fshift = dx10.0_dy0; Fspeed = s35;
direct # 'head';
Fp = {[+'link'; ^ Fp];
     SPEED = Fspeed; WHERE = Fshift};
repeat(^ Fp)
```

COORDINATED MOVEMENT OF A GROUP

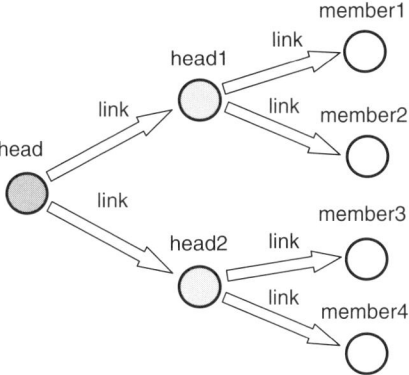

Figure 6.3. Three-layer hierarchy.

This program repetitively uses recursive procedure Fp providing top-down navigation of the hierarchy with bottom-up acknowledgment of actions in nodes (rule none—its bracketed option—simplifies the distributed control, allowing us to avoid the use of rule sequence).

For example, we can create a three-layer hierarchy shown in Figure 6.3, where an additional layer to the two-layer hierarchy in Figure 6.2 has two local heads: head1 and head2 (subordinate to global head), and each local head has two subordinate group members.

The program creating this hierarchy (also supplying certain initial coordinates to the nodes) will be as follows:

```
create(
  direct # 'head'_x25.5 y40.2;
  (+'link' # 'head1'_x35.5 y30.2;
    +link # ('member1'_x45.4 y25.2, 'member2'_x45.4 y35.2)),
  (+'link' # 'head2'_x35.5 y50.2;
    +link # ('member3'_x45.4 y45.2, 'member4'_x45.4 y55.2))
)
```

A stepwise movement in space of the hierarchy created by this program after applying to it the previous, recursive, program is shown in Figure 6.4.

6.1.4 Simultaneous Movement of All Nodes in a Group

Any other forms of movement of hierarchically organized groups can be possible too, for example, simultaneous (not stepwise, as before) movement of the whole group, both its heads on different layers and fringe members, while obeying the top-down chain of

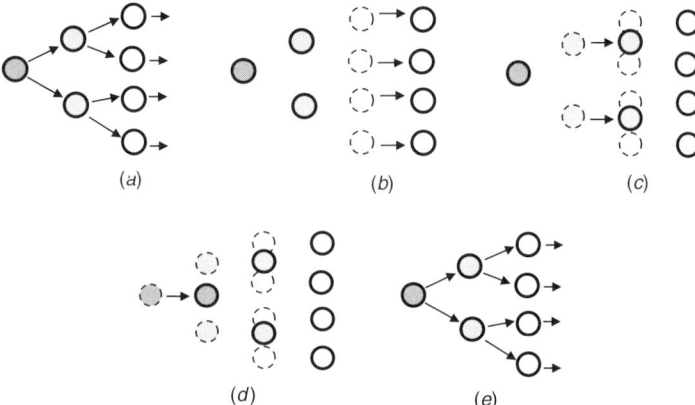

Figure 6.4. Coordinated movement of a three-level hierarchy: (a) top-down spread of global command from the head, (b) movement of fringe nodes, (c) movement of second-level nodes, (d) movement of the head, and (e) same as in (a).

command, as follows (by just changing the position of square brackets in the previous group-moving program!):

```
Fshift = dx10.0_dy0; Fspeed = s35;
direct # 'head';
Fp = {(+'link'; ^ Fp),
      (SPEED = Fspeed; WHERE = Fshift)};
repeat([^ Fp])
```

During the recursive descent of control through the hierarchy, each node starts moving immediately, passing the control further down simultaneously with its own movement, and not waiting for the replies from subordinates about the completion of their steps.

6.1.5 Moving to the Averaged Positions of Subordinates

We can complicate the group movement task in a way where the movement of nodes should not always be by the predetermined coordinate shift, but rather to the averaged center among actual coordinates reached by the node's direct subordinates. And only the fringe nodes of the hierarchy should use the predetermined shift to move in space (as having no subordinate nodes). The WAVE-WP program for this modification will be as follows:

```
Fshift = dx10.0 dy0; Fspeed = s35;
direct # 'head';
Fp = {New = average(+'link'; ^ Fp); SPEED = Fspeed;
      WHERE = or((New != nil; New), Fshift); WHERE};
repeat(^ Fp)
```

The movement of heads for this modification can also be made simultaneous with the movement of subordinate nodes. The heads can use previous averaged coordinates of the subordinates or the initially set up coordinate shift for all nodes (in Fshift), if no coordinates from the subordinates have been received yet, as follows (by a minimal modification of the previous program—just changing a semicolon to comma in one place, and adding braces and square brackets in other places!):

```
Fshift = dx10.0 dy0; Fspeed = s35;
direct # 'head';
Fp = {New = average(+'link'; ^ Fp),
     (SPEED = Fspeed;
      WHERE = or((New != nil; New), Fshift); WHERE)};
repeat([^ Fp])
```

6.1.6 Further Possible Group Movement Modifications

In further modifications, the top head of the group can also change the global direction of movement of the group by changing the value in Fshift at any time, making this decision, say, on the basis of global awareness of the distributed situations and in pursuit of established goals (and these can be added as another, distributed, payload to the previous programs).

We can set up at the beginning a sequence of coordinate shifts, which should be applied during the group movement, instead of the single shift applied repetitively, as before. And instead of a sequence of shifts, there can be a sequence of waypoints with absolute coordinates, which should be followed by the top head, and which can be easily converted into common shifts to be used by all other nodes, wherever they are located. These shifts can be broadcast down the hierarchy to all participants. The shifts can also be delivered to fringe nodes only, and then the higher levels can go to the averaged positions obtained after the move of subordinate nodes, as in the previous examples.

In WAVE-WP, we can also organize both local and global analysis of the spatial shape of the group (which may be three rather than two dimensional as before, say, for unmanned air or underwater groups) and correction of individual node movements, to restore and support the original shape during the distributed propagation in space, and so on.

6.1.7 Reverse or Heads-First Movement

So far we considered the hierarchically organized group movement where the subordinate entities were moving first, or at least simultaneously, with the higher levels. This order of movement may be suitable, for example, for combat situations, where higher levels of hierarchy should be better protected from possible damages and should move only after reconnaissance is made by the lower levels, and into the cleared positions.

A reverse order, where higher levels are moving first and the lower levels afterwards, while preserving the group's spatial shape, may be of importance too, for example, on a march to new battle locations—of course, if conditions (say, terrain) permit such a distributed movement.

After moving of the top head by the given value in Fshift, the actual shift for subordinate nodes may be easily calculated from the new and previous positions of the head node and sent down the hierarchy to the subordinate nodes, for them to move in space too. The latter nodes will do the same for their own subordinates, and so on.

A corresponding solution in WAVE-WP for this group "heads-first" movement scenario follows (assuming that the group moves opposite to the previously used direction, i.e., from right to left now, which is accounted in the new initial content of Fshift):

```
Fshift = dx-10.0 y0; Fspeed = s35;
direct # 'head';
repeat([repeat(Ncurrent = WHERE; WHERE += Fshift;
              Fshift = WHERE - Ncurrent; +'link' # any)])
```

Using this program, the movement of the three-level hierarchy (depicted in Fig. 6.3) will take place as shown in Figure 6.5.

6.1.8 Movement in a Column

Another movement of a group with the heads-first mode may be in a column, say, on a road, where the presented above two-dimensional distributed movement on a plane cannot be possible in principle. If the hierarchy is just a line, not a tree, we can use the previous program for movement without any changes.

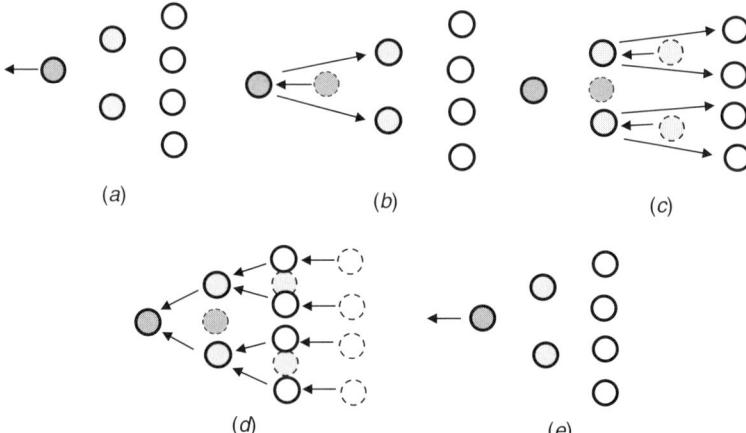

Figure 6.5. Coordinated movement with heads first: (a) starting position, with the head to move, (b) movement of the head and passing its updated coordinate shift to the second level, (c) movement of the second-level nodes and passing updated coordinate shifts to the third level, (d) movement of the third level and echoing termination upwards the hierarchy, and (e) same as in (a).

COORDINATED MOVEMENT OF A GROUP

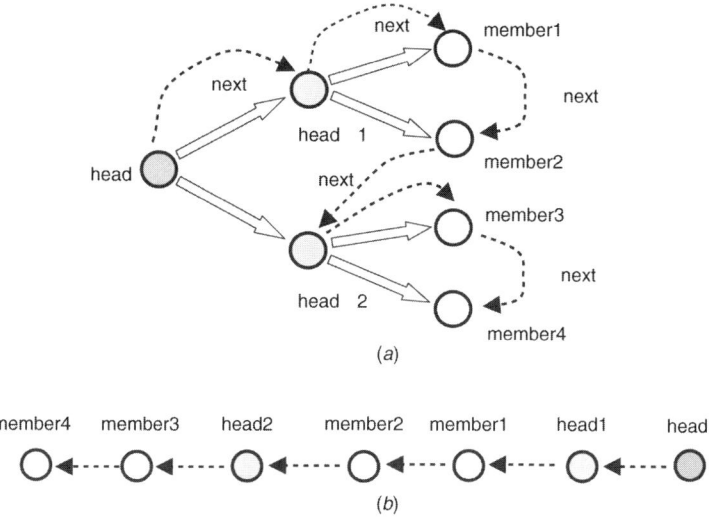

Figure 6.6. Establishing the column infrastructure: (a) setting additional oriented links next and (b) initial redeployment of the column for movement in PW.

But if the group infrastructure is not set up yet, or is a tree, as in Figures 2.2 or 6.3, in addition to it we can establish a special column infrastructure (defining the order of movement and special command-and-control channels), using additional oriented virtual links (let them be named next, as shown in Fig. 6.6a). For a real movement as a column, we should also physically redeploy all nodes of the group to form a line, as shown in Figure 6.6b.

This adding of new, next, links and physically redeploying the existing group nodes to form a line in PW can be made by a single program, which may look as follows (using, for short, the allowed abbreviation W for the environmental variable WHERE):

```
create(
  direct ## 'head'; W = x90 y70; +'next' ## 'head1';
  W = x80 y70; +'next' ## 'member1'; W = x70 y70;
  +'next' ## 'member2'; W = x60 y70; +'next' ## 'head2';
  W = x50 y70; +'next' ## 'member3'; W = x40 y70;
  +'next' ## 'member4'; W = x30 y70
)
```

The repetitive program shown above for the movement of a multilevel hierarchy can be applied for an arbitrary column too, after changing the name of links to follow and setting a proper coordinate shift for each step:

```
Fshift = dx5.0_dy0; Fspeed = s35;
direct # 'head';
repeat([(repeat(Ncurrent = WHERE; WHERE = Fshift;
                Fshift = WHERE - Ncurrent; +'next')])
```

166 EXEMPLARY MISSION SCENARIOS

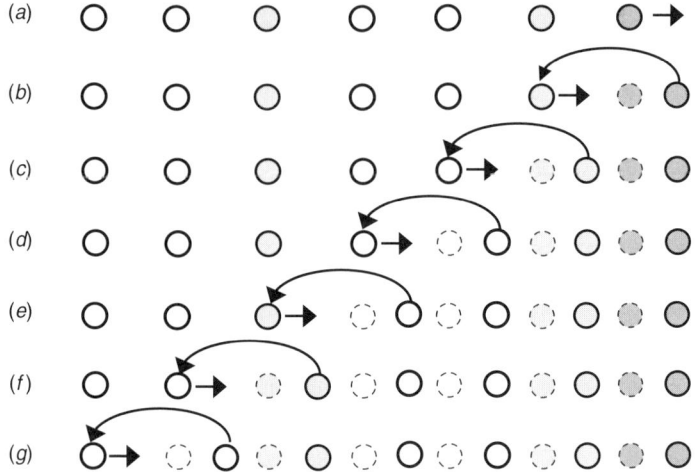

Figure 6.7. Stepwise movement of nodes of the column with transference of control.

A stepwise movement of the column using this program, with the transference of control between nodes is depicted in Figure 6.7a–g.

6.1.9 Integrating Different Movement Solutions

We can easily integrate within one simple program in WAVE-WP different forms of group movement discussed above—say, movement of a hierarchy with any number of levels and any number of nodes and a column of any length, allowing fringe nodes to move first or heads move first (respectively, tail first or head first for a column). Such a universal program using, for simplification, the same constant shift for all nodes and for all steps of their movement will be as follows:

```
Fshift = ...; Fspeed = ...; Flink = ...; Fmode = ...;
direct # 'head';
Fp = (
  {[+ Flink # any; ^ Fp:1]; SPEED = Fspeed; WHERE = Fshift},
  {[repeat(SPEED = Fspeed; WHERE = Fshift; + Flink # any)]}
);
repeat(^ Fp : Fmode)
```

This program may be applied from the head of any hierarchical or column infrastructures, previously setting up coordinate shift for any movement in Fshift, speed of movement in Fspeed, particular infrastructure by naming proper links in Flink (i.e., link for a hierarchy and next for a column), and mode of movement in Fmode (1 for fringe or tail nodes moving first and 2 for heads moving first).

6.2 PHYSICAL MATTER DELIVERY AND REMOTE PROCESSING

The general task here is a delivery of a certain volume of substance from a starting point to a target, with processing it at the latter, as shown in Figure 6.8. The substance may, for example, be fire-extinguishing chemicals, and the destination may reflect the peak of intensity of a forest fire, say, on top of the hill. (Any other interpretations of this scenario can also be possible.) Different levels of the solution of this task can be organized in WAVE-WP.

6.2.1 Most General Task Solution

We will start with the most general description of the problem, just following the definition given above and shown in Figure 6.8, for which the wave program will be as follows:

```
direct # x-2.3 y0.7;
Fload = "60 kg chemicals";
direct # x0.9 y0.5; Fload ? apply
```

In this program, a hop in space will be performed with the given x and y coordinates as -2.3 and 0.7 for the start and 0.9 and 0.5 for the target. The 60 kg substance is lifted at the start and assigned to mobile variable Fload, which will be subsequently processed by the external procedure apply at the destination.

6.2.2 Splitting into Subtasks

Imagine that the delivery of physical matter may not be possible by usual transportation means on the whole way from the start to the target. Say, the area around the target (as the center of fire in our case) is too dangerous to be passed by a usual vehicle (ground or aerial), and for the final part of the journey the matter should use some express means, say, a shell to be fired at some distance from the target.

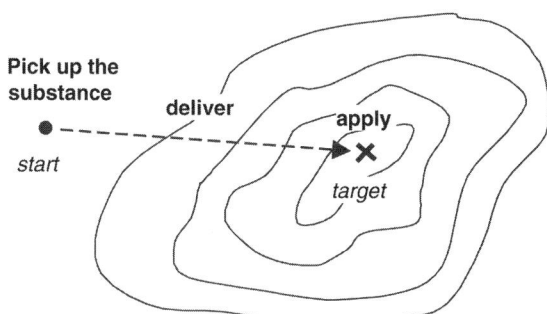

Figure 6.8. Most general task definition of physical matter delivery and processing.

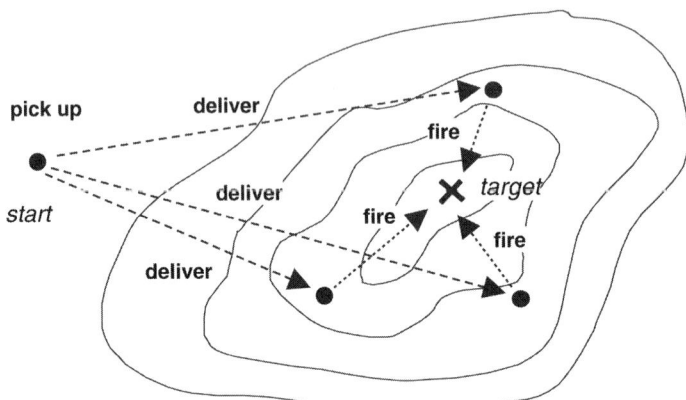

Figure 6.9. Splitting into parallel subtasks, with independent firing.

As there may be limitations on the shell's weight, the whole matter should be split into portions and put into a number of shells. Also, for safety and for improving an impact on the target, the shells should be fired in parallel from different positions around the target, being delivered to these positions independently from the starting point, as shown in Figure 6.9.

The program for this scenario, with the chemicals put into 20-kg shells assigned to the replicating mobile variable Fload for the delivery into three firing positions with certain coordinates, will be as follows (it is assumed that the variable's replication is allowed by sufficient physical resources at the starting point, i.e., the shells):

```
direct # x-2.3 y0.7; Fload = "20 kg shell";
direct # (x1.2 y1.2, x-0.1 y-0.3, x1.4 y-0.3);
(Fload, x0.9 y0.5) ? fire
```

The shells will be fired independently from these positions by the procedure fire in each of them, using the same target coordinates as a parameter.

6.2.3 Adding Synchronization

Imagine that for a further strengthening of the impact on the target, all the three fire-extinguishing shells should be fired simultaneously after being delivered to the firing positions. This can be achieved by using rule synchronize, as follows:

```
direct # x-2.3 y0.7; Fload = "20 kg shell";
synchronize(direct # (x1.2 y1.2, x-0.1 y-0.3, x1.4 y-0.3));
(Fload, x0.9 y0.5) ? fire
```

6.2.4 Setting Specific Routes

To add further details to the above scenario, let us consider the reaching of each firing point via a sequence of some intermediate waypoints, before the synchronized firing takes place, as shown in Figure 6.10.

The program creating three repeating space propagation branches (using rule `repeat`), with the route coordinates kept in variable `Fpoints` (individually for each branch), will be as follows:

```
direct # x-2.3 y0.7; Fload = "20 kg shell";
synchronize(
  Fpoints = (x-1.2 y1.0, x0.1 y1.2, x1.2 y1.1),
  Fpoints = (x-1.0 y0.1, x-0.1 y-0.3),
  Fpoints = (x-1.5 y-0.5, x0.1 y-0.9, x1.4 y-0.3);
  repeat(
    Fpoints != nil; direct # Fpoints : 1;
    Fpoints : 1 = nil
  )
);
(Fload, x0.9 y0.5) ? fire
```

In this program, where the used waypoints are being removed from their sequences, the firing action is synchronized upon reaching the firing positions (the latter being the last waypoints of each route).

6.2.5 Assigning Robots to Scenarios in Column Movement

The scenarios shown above are written as integral programs reflecting the semantics of the tasks to be solved, rather than directives to robots on what and how they should do

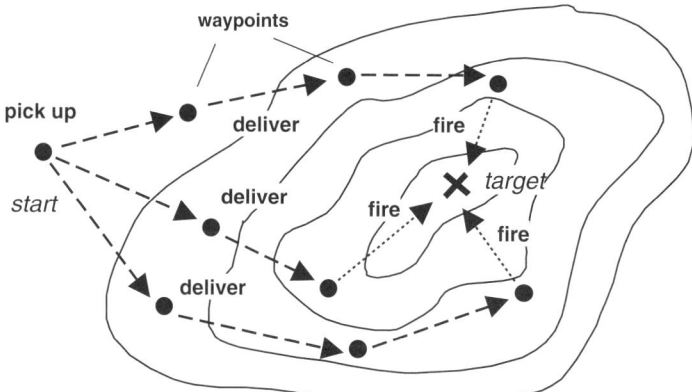

Figure 6.10. Setting specific routes.

and how to coordinate activities. This allows us to abstract from the implementation details and concentrate on the mission goals, strategy, and tactics instead. Assignment of robots to these scenarios can be done automatically. These scenarios can also be executed by a different number of robots, and the failed robots can be substituted at run time by other robots automatically.

However, if needed, the mission scenarios can also be written on the level of a detailed movement of robots and cooperation between them, as usual. For this, let us consider a program where each shell is delivered to its firing point by a separate mobile robot, and the robots move to the firing positions together in a column in the established order, strictly following each other. Only the first, or head, robot can know the whole route, sending exact coordinates of the reached points to the next robot (which does the same for its own follower in the column, and so on).

Let us also assume, for simplicity, that the robots in the column are synchronized only after passing the updated waypoints, and the firing positions are the last three waypoints of the column's route. These are to be finally occupied by the robots in the order they keep in the column, as shown in Figure 6.11. Upon reaching these positions, the robots should shell the target simultaneously, in a synchronized way.

The following program describes this scenario in detail. It first establishes a column-oriented command-and-control infrastructure between robots, using the rule create, linking the robots by oriented virtual links (named next), with + and − subsequently reflecting their traversing directions in command-and-control operations. It also places the robots in proper starting positions, allowing them to keep some starting distance from each other, before they deploy themselves into a moving column operating via the created infrastructure.

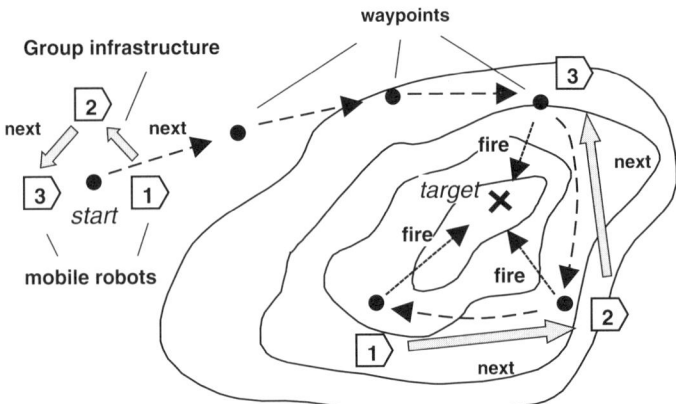

Figure 6.11. Performing the task by an organized column of robots.

```
sequence(
  create(
    direct ## robot.1; WHERE = x-1.9 y0.8;
    +'next' ## robot.2; WHERE = x-2.5 y1.3;
    +'next' ## robot.3; WHERE = x-2.8 y0.6
  ),
  release(
    direct # (robot.2, robot.3); Nammun = "20 kg shell";
    repeat(
      New != 'finish'; wait(New != nil); N = New; New = nil;
      (+'next' #; New) = WHERE; WHERE = N
    ); New = nil
  ),
  release(
    direct # robot.1; Nammun = "20 kg shell";
    Npoints = (x-1.0 y1.1, x0.1 y1.4, x1.2 y1.2,
               x1.4 y-0.3, x-0.1 y-0.3);
    repeat(
      Npoints != nil; (+'next' # any; New) = WHERE;
      WHERE = Npoints : 1; Npoints : 1 = nil
    );
    repeat(+'next' #; New = 'finish'; wait(New != nil));
    repeat(release((Nammun, x0.9 y0.5) ? fire); -'next' #)
  )
)
```

The created virtual infrastructure is used by robots for passing the exact, updated, coordinates of the reached waypoints to the subsequent robots, and in the end—for waiting for the arrival of all of them into their firing positions, with subsequent synchronized firing. The latter two operations are initiated by the head robot, upon arrival into its own firing position. For these operations, propagation through the column's distributed infrastructure takes place in both directions.

As can be seen from the above program examples, we can describe problems to be performed by groups of robots on quite different levels of abstraction. These levels may range from fully semantic descriptions of what to do in principle in the distributed space to the detailed tasking of individual robots, and establishing their interactions and synchronization. All these scenarios are effectively expressed within the same simple WAVE-WP formalism.

6.3 PHYSICAL WORLD SEARCH ASSISTED BY VIRTUAL WORLD

Many problems potentially requiring distributed robotics are linked with the parallel search of large physical spaces. De-mining territories, looking for minerals (domestically or on other planets), patrolling regions, cleaning territories, robotic fishery, and disaster relief may be a few of them. Problems such as de-mining, for example, may need an absolute guarantee that the whole territory is thoroughly examined.

One of the main problems here is that it is difficult, often impossible, to keep a central database of the space to be searched. This is because most of the information is highly dynamic and changeable at run time and can be updated in local brains of robots only, which move through and see only small parts of the space.

Also, in dynamic and hostile environments, it is impossible to make reliable optimization and scheduling in advance, as the time of space search and processing may depend on emergent circumstances, and individual robots may fail indiscriminately at run time.

We propose here a general solution of this problem based on WAVE-WP, where a team of robots can effectively convert into a universal spatial computer working without any central resources and capable of solving any space search and optimization problems autonomously, in a fully distributed and parallel manner.

6.3.1 Creating the Distributed Virtual World

We will be using the traditional representation of a two-dimensional space by the set of polygons, as shown in Figure 6.12, where each polygon keeps parameters describing its dimensions and shape, sufficient for navigation in the corresponding part of PW.

The set of data describing polygon details, together with their connectivity graph (CG) can be represented in WAVE-WP in an integral form as a distributed knowledge network (KN), shown in Figure 6.13. The polygon data in this representation will exist as nodes pi, these nodes being attached to the corresponding CG nodes i by links named p, with g used as the name of all CG links.

A variety of possibilities may be offered in WAVE-WP for the efficient expression and creation of such a network in the distributed environment. Most compact ones use a parallel depth-first spanning tree template, with an example shown in Figure 6.14 by bold oriented arcs (if starting from node 1), with the network creation steps numbered at the arcs (the same numbers identifying potentially parallel steps).

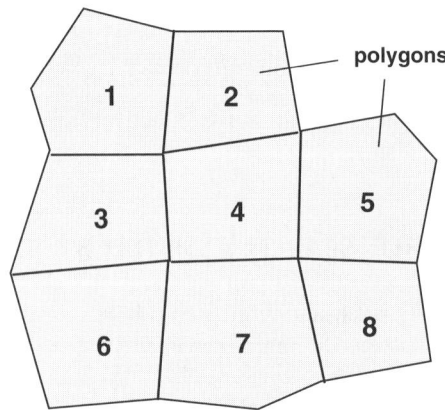

Figure 6.12. Linked polygons representing the space to be searched.

PHYSICAL WORLD SEARCH   173

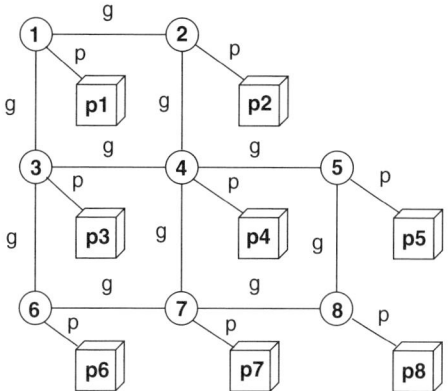

Figure 6.13. Representing the space as a knowledge network.

The WAVE-WP program for creating this network using the depth-first spanning tree template shown in Figure 6.14 will be as follows:

```
create(
  direct#1; [p#p1]; g#2; [p#p2]; g#4; [p#p4];
  g#7; p#p7, (g#6; [p#p6]; g#3; p#p3, g##(1, 4)),
  (g#8; [p#p8]; g#5; p#p5, g##4)
)
```

It is supposed that pi in this program are not the node names but rather stand for polygon information, which should substitute them for real applications. This KN creation program can start from any robot, and the KN will be automatically distributed

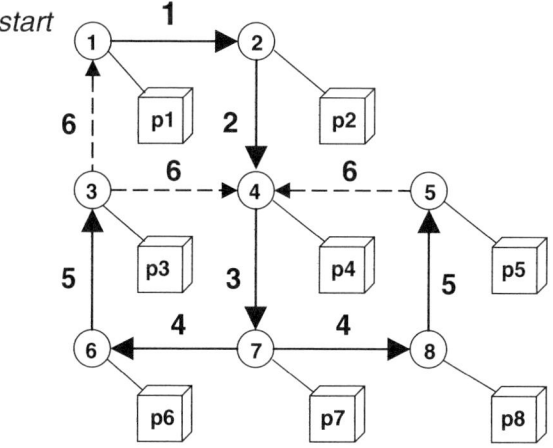

Figure 6.14. Depth-first spanning tree template.

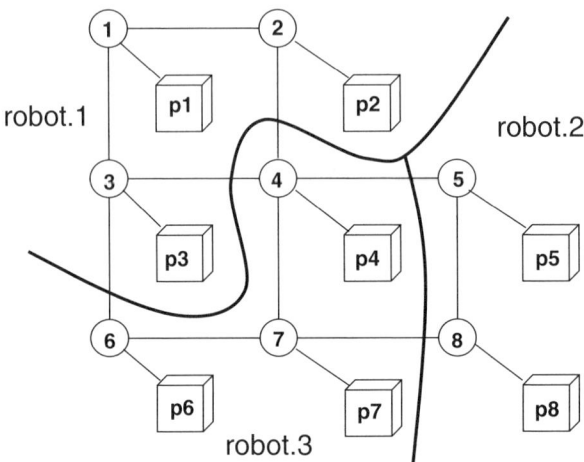

Figure 6.15. Network distribution between robots.

between their artificial brains in an optimized way, where some links may happen to connect nodes in different robots. If needed, the distribution between particular robots can be set up explicitly, as follows (abbreviation R for the environmental variable RESOURCES is used, for short):

```
create(
  R=robot.1; direct#1; [p#p1]; g#2; [p#p2];
  R=robot.3; g#4; [p#p4]; g#7; p#p7,
  (g#6; [p#p6]; R=robot.1; g#3; p#p3, g##(1, 4)),
  (R=robot.2; g#8; [p#p8]; g#5; p#p5, g##4)
)
```

The resultant distribution of KN between the robots is shown in Figure 6.15.

A variety of space search and processing tasks can be effectively solved in this virtual world, represented as KN, in parallel and fully distributed modes. The solutions found on the KN cooperatively can be subsequently or simultaneously converted into physical movement of robots, and these phases may overlap in time and space as one integral process. We will consider hereinafter only elementary programming examples, with orientation on space-cleaning applications, using different levels of abstraction.

6.3.2 Top-Level Space-Cleaning Scenarios

On the top level, we may write in WAVE-WP just what we generally want: To reach and clean all polygons simultaneously and, say, only once. The corresponding program will be as follows:

```
direct # (1, 2, 3, 4, 5, 6, 7, 8); (p #) ? move_clean
```

PHYSICAL WORLD SEARCH

The invoked external aggregated procedure move_clean will use the polygon parameters (accessed from the corresponding CG nodes by links p) to move physically into the PW regions corresponding to these polygons (while formally staying in the virtual CG nodes) and then clean these regions.

Another desire may be to have a simultaneous and repeated cleaning of every polygon with some interval, say an hour, while staying in the polygons indefinitely after reaching them only once. Using environmental variable TIME, this will look as follows:

```
direct # (1, 2, 3, 4, 5, 6, 7, 8); (p #) ? move;
repeat((p #) ? clean; TIME += 360)
```

If we would want a repeated simultaneous cleaning of all polygons with revisiting them each time (say, using the robotic resources elsewhere during the breaks), the program will be as follows:

```
repeat(
  [direct # (1, 2, 3, 4, 5, 6, 7, 8); (p #) ? move_clean];
  TIME += 360
)
```

The distributed brain of the robotic group will be dynamically appointing available robots to polygons, so actual parallelism will depend on the dynamic availability of mobile hardware resources. The efficiency of the shared use of robots will depend on the internal WAVE-WP distributed interpretation and optimization mechanisms, which may not always be optimal if high-level scenarios are used with minimum details.

6.3.3 Single-Step Multiple-Branch Search

We may, however, use compromise solutions and assist the distributed WAVE-WP interpreter in making hardware distribution decisions (especially if the hardware resources are limited) by providing a more detailed description of the problem, say, as a limited set of competing or cooperating branches—but still well above the implementation layer. Each branch will move between the noncleaned polygons and clean them, while blocking other branches from visiting the occupied polygons.

Visiting Polygons Only Once. With visiting polygons only once, and moving to neighboring polygons only, while checking the states of surrounding neighbors sequentially, the program, consisting of three competing branches starting in polygons 1, 4, and 7, will be as follows:

```
direct # (1, 4, 7); M = 1; (p #) ? move_clean;
repeat(
    or(g #; grasp(M == nil; M = 1)); (p #) ? move_clean
)
```

where nodal variable M with value 1 marks polygons as cleaned in the corresponding CG nodes. Each branch terminates if there are no neighboring noncleaned polygons.

This solution, based on searching neighboring noncleaned polygons only, may, however, fail to clean all polygons, as the remaining noncleaned polygons may happen to be separated from the current positions in space by the cleaned polygons.

Regular Random Search. Another variant with local search only may be regular random movement to empty neighboring polygons, with repeated visiting and cleaning of the same polygons allowed, and infinite attempts to find an empty neighbor, as follows:

```
direct # (1, 4, 7); M = 1; (p #) ? move_clean;
repeat(
  or(
  random(
    g #; grasp(M == nil; M = 1);
    (BACK #; M) = nil; (p #) ? move_clean),
  nil))
```

If the cleaning of polygons is required only once, with revisiting of the same polygons still allowed, the previous program can be easily modified for this as follows (using additional nodal variable N saving the state of the polygon with which it was entered):

```
direct # (1, 4, 7); M = 2; (p #) ? move_clean;
repeat(
  or(
    random(
      g #; grasp(M != 2; N = M; M = 2); (BACK #; M) = 1;
      (p #) ? move; [N == nil; (p #) ? clean]),
    nil))
```

The above two solutions guarantee that all polygons will sooner or later be cleaned.

Global Search for the Next Polygon. We may also introduce a more expensive global search for a noncleaned empty polygon, not necessarily neighboring one, by accessing the whole KN, which will always guarantee a full solution of the problem. Robots may, however, need to move to the destination polygons through the occupied polygons, so collision avoidance mechanisms must be involved on the implementation level. The program code for this case will be as follows:

```
direct # (1, 4, 7); M = 1; (p #) ? move_clean;
repeat(
  or(
    direct # any; (g #) != nil;
    grasp(M == nil; M = 1); (p #) ? move_clean))
```

In a sequentialized hop (to reduce competition between the tree branches) to all other nodes by each branch, only CG nodes are of interest (being distinguished from polygon nodes by having incident g-type links).

Combined Local–Global Search Solution. A combination of local and global search, where global search is invoked only if local search fails, may be particularly useful, leading to a full solution of the cleaning problem, which is more economical than the pure global search. The following program uses search of neighbors from one of the earlier examples and global search from the previous example, terminating if global search, invoked after the failed local search fails to find an empty noncleaned polygon:

```
direct # (1, 4, 7); M = 1; (p #) ? move_clean;
repeat(
  or(
    or(g #; grasp(M == nil; M = 1)),
    or(direct # any; (g #) != nil;
      grasp(M == nil; M = 1));
    (p #) ? move_clean)
)
```

Assignment of Robots to Branches. The assignment of robots to branches will be done by the distributed interpreter automatically. But we may also assign particular robots to particular branches explicitly, using environmental variable RESOURCES (or R, for short) and modifying the previous program as follows:

```
(direct # 1; R = robot.3), (direct # 4; R = robot.2),
(direct # 7; R = robot.1); M = 1; (p #) ? move_clean;
repeat(
  or(
    or(g #; grasp(M == nil; M = 1)),
    or(direct # any; (g #) != nil;
      grasp(M == nil; M = 1));
    (p #) ? move_clean)
)
```

Keeping the robot's name in environmental variable RESOURCES for different branches, each branch will always drive through physical space only the robot dedicated to it despite, possibly, propagating electronically through other robot's brains holding the currently navigated virtual world partitions.

The starting distribution of robots between the polygons and initial cleaning activity in them by this program is shown in Figure 6.16. The occupation of polygons is reflected by shading their CG nodes, with cleaned polygons having their nodes (boxes) shaded too.

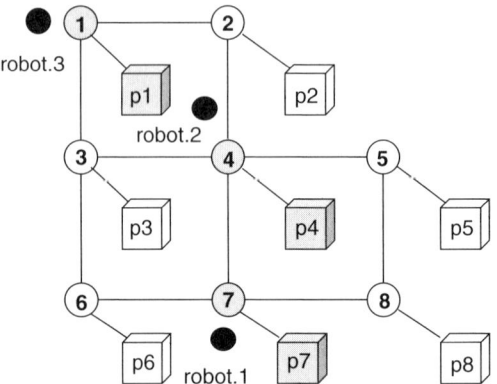

Figure 6.16. Initial distribution of robots.

6.3.4 Full-Depth Search for Polygons

We have considered a single-depth search of CG for empty polygons (either via surface links to direct neighbors or directly to any other nodes by tunnel hops). Much better solutions can be achieved by a nonlocal, up to full depth, search of CG for each branch via surface links, with finding optimum paths to empty and noncleaned polygons.

The following program first creates a breadth-first spanning tree of CG from a position occupied by the robot, avoiding occupied polygons and terminating in the first met noncleaned and empty polygons. Then the closest such polygon is chosen to be the next one to clean, and the found path to it is used to physically drive the robot to the destination, possibly, through other (empty only) polygons.

```
(direct # 1; R = robot.3), (direct # 4; R = robot.2),
(direct # 7; R = robot.1);
PAINT = R; M = 2; (p #) ? move_clean; N=1;
repeat(
  F = min(
    repeat(
      g #; grasp(M != 2; N == nil; N = 1);
      Fpath _= CONTENT; Fdistance += 1;
      or((M == nil; Fdistance _ Fpath ! done), nil)
    )
  );
  F != nil; N = nil; M = 1;
  repeat(F:1 = nil; F != nil; g # F:1; (p #) ? move);
  M = 2; (p #) ? clean
)
```

This program, for example, will find paths 2-4-7-8 and 2-4-3 to the first met noncleaned polygons p8 and p3 for robot 3 staying in p2, as shown in Figure 6.17,

PHYSICAL WORLD SEARCH

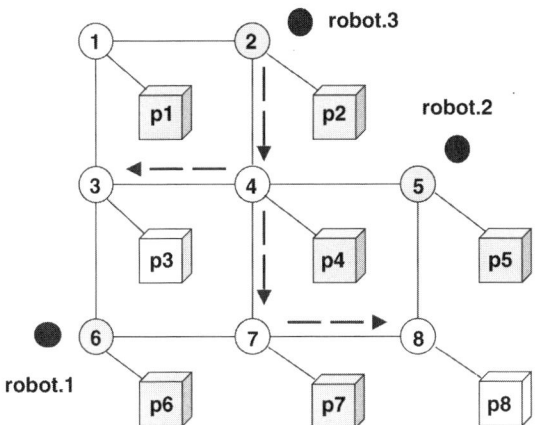

Figure 6.17. Spanning tree search for unvisited polygons.

choosing polygon p3 to be cleaned next as having a smaller distance from p2 with robot.3 (two versus three for p8). The program will subsequently drive robot.3 to the area in the physical world corresponding to polygon p3, via the waypoints found (here polygon p4).

This may not be a full solution, however. But if we allow the spanning tree to develop via occupied polygons too, not only free ones, the solution of finding paths to noncleaned polygons will become universal, with the potential possibility of collisions, however. The above program can do this with elementary modifications, resulting in treating, for example, the path 2-4-5-8 for robot 3 (see Fig. 6.17) as legal too. We can also easily express in WAVE-WP a balanced universal shortest path tree solution, where moving through occupied polygons essentially increases the weight of the path. So the paths via empty polygons may be more preferable, if they exist in principle.

6.3.5 Run Time Space Modification

The proposed organization of multirobot space search may be dynamic and open, with new search regions and new robots added at run time, while others are removed in a seamless way. CG and polygons can be redistributed between robots at run time, say, when certain robots are overloaded or fail. For example, adding new adjacent polygons p9 and p10 and linking them with existing adjacent ones p7 and p8, correspondingly, via new CG nodes 9 and 10, can be done by the following program:

```
create(direct # 9; [p # p9, g ##7]; g # 10; p # p10, g ## 8)
```

If we want this new network part to be placed in a particular robot, we should write:

```
R = robot.4;
create(direct # 9; [p # p9, g ##7]; g # 10; p # p10, g ## 8)
```

The KN nodes can move freely between robots in WAVE-WP. For example, we can easily redeploy nodes 1 and 3 from `robot.1` to `robot.3`, node 4 from `robot.3` to `robot.1`, and node 7 from `robot.3` to `robot.2`. Links from these nodes to other nodes as well as between them will be saved during the movements. These changes can be done by the following program:

```
(direct # (1, 3); DOER = robot.3), (direct # 4; DOER = robot.1),
(direct # 7; DOER = robot.2)
```

We can also remove existing polygons at run time such as, say, p6 together with the corresponding CG node 6, as follows:

```
direct # 6; p #, nil; CONTENT = nil
```

All these modifications can be performed separately, one after the other, or simultaneously, at any time, starting from the same or from different robots. The resultant distribution of KN between the robots taking into account the changes presented above, including the new nodes added, is shown in Figure 6.18.

The changes in the distributed knowledge network may also take place simultaneously with any wave processes on this network. For example, all the space

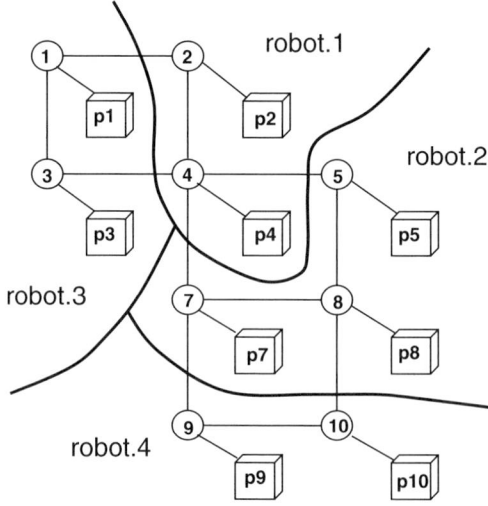

Figure 6.18. Resultant distribution of the modified virtual world.

search and cleaning solutions presented above will work during the run time modification of the virtual space on which they operate.

6.4 MAP-BASED COLLECTION OF SAMPLES

This final scenario for the current chapter assumes that a certain area on the planet surface must be simultaneously navigated by multiple rovers, all starting from a common landing point. The rovers should collect soil samples at the points given and return them all to the lander.

The surface navigation is supposed to be assisted by the territory map reflecting points of interest and their accessibility for each other, represented altogether as a knowledge network, KN. Its nodes keep the corresponding PW point coordinates (also names—just numbers, for simplicity), being of pv types. KN links, reflecting suitable paths between the points, are marked with expected propagation time rather than distance (this time based, say, on known peculiarities of the terrain), as shown in Figure 6.19.

This simplified map of the region of interest can be created and kept in a fully distributed form by the following wave program, which is structured as a possible depth-first spanning tree template of the KN:

```
create(
  direct # 5_x60y45; 20 # 1_x30y45; 15 # 2_x55y65;
  [9 ## 5]; 5 # 3_x125y70; [3 # 4_x100y55; 4 ## 2];
  1 # 8_x140y30; 5 # 7_x115y30; 3 # 6_x65y20;
  10 ## 1, 8 ## 5
)
```

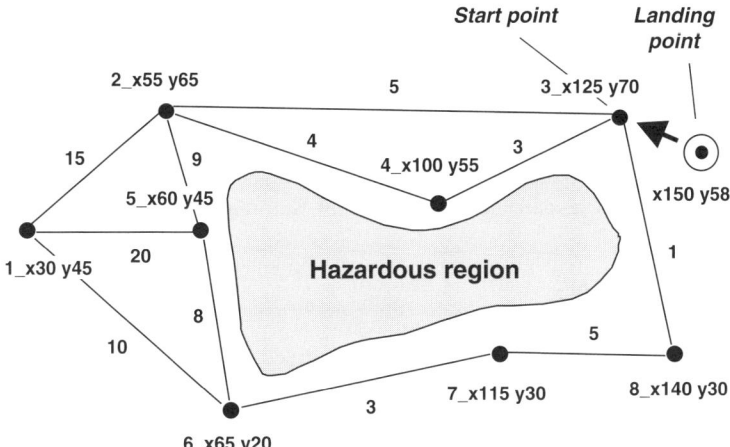

Figure 6.19. Simplified territory map as a knowledge network.

The actual landing place in the region of interest may depend on different circumstances (e.g., weather), so after the landing, a proper starting point on the map should be chosen, and then optimal access routes to other points should be found via the map, for a subsequent physical navigation of the region.

The following parallel program, applied at the landing site, first finds the nearest point on the map from the landing, and then creates a shortest path tree (SPT) from this point to all other points, taking into account the KN topology and weights of links between its nodes. The program subsequently navigates the planet surface physically, being guided by the SPT found and using the recursive procedure Fdown, gradually mapping the virtualized SPT nodes onto the corresponding PW locations.

Soil samples are picked up in all the reached points (using external procedure liftSample) and returned to the landing site via the optimal routes defined by the SPT (this is done automatically, on the internal interpretation level). All these samples are being ultimately stored in the PW-bound variable Nsamples at the landing site (this variable holding physical matter rather than information).

```
Flanding = WHERE;
Nstart = min(
  direct # any; (WHERE, Flanding) ? distance_ADDRESS): 2;
Nsamples = (
  direct # Nstart;
  [repeat(
    any # any; Fdistance += LINK;
    grasp(
      or(Ndistance == nil, Ndistance > Fdistance);
      Ndistance = Fdistance; Npredecessor = BACK
    )
  )];
  Fdown = {
  ? liftSample,
  (+ any # any; Npredecessor == BACK; ^ Fdown)};
  ^ Fdown
)
```

The simplicity and compactness of the program above can be explained by the fact that it describes semantics of the problem to be solved directly, without implementation details. Dynamic optimized mapping of this high-level spatial scenario onto available mobile hardware can be carried out in a formal, automatic, and fully distributed way.

6.5 CONCLUSIONS

A variety of exemplary spatial scenarios expressed in WAVE-WP have been considered, analyzed, and discussed. These scenarios have the following common features:

- They describe free movement and operation in the distributed physical world directly.

CONCLUSIONS

- They are using the virtual world, represented as the knowledge network, and navigation in it to guide the movement and operation in the physical world.
- Dynamic task-oriented virtual infrastructures are effectively created and maintained within these scenarios, simplifying hierarchical command and control and pursuit of local and global goals by robotic groups.
- The scenarios express high potential parallelism of operations and fully distributed solutions in both physical and virtual worlds, as well as in their integration, without any central resources.
- The obtained integral spatial solutions are not divided into agents with their interaction and synchronization, as usual, as these are effectively shifted to the automatic implementation layer. This dramatically reduces the complexity and length of the application code.
- The same simple WAVE-WP formalism is used for the description of spatial solutions on the most different layers—from pure semantic (of just what to do in distributed space) to the most detailed ones (of how to do it, by which technical means, and how these means should operate and interact).
- The same scenarios can include solutions with a combination of different layers, behaving altogether as a single spatial agentless program, without any seams between its parts or parts of the distributed virtual networks, created, navigated, and processed by these scenarios.

7

DISTRIBUTED MANAGEMENT USING DYNAMIC INFRASTRUCTURES

As world dynamics is increasing rapidly (military conflicts, terrorist attacks, consequences of global warming, etc.), crisis management is becoming one of the hottest topics in both civil and military areas. Rapid composition of an efficient crisis reaction force (or CRF), whether national or multinational and providing its high operability and survivability in unpredictable and hostile environments, is an extremely important and complex problem to be solved.

Advanced CRF should efficiently work in manned, automated, and, possibly, fully automatic mode, with the use of (massive and cooperative) robotics within the force mix and unified command-and-control infrastructures. Such a force, possibly, composed from dissimilar and casual components, can be heterogeneous and distributed over a large territory. It should, however, always remain operational and manageable as a whole, capable of pursuing local and global goals and assessing distributed situations, despite the high probability of indiscriminate damages of its components and communications between them.

Infrastructure of a system is usually considered as something solid, static—as the backbone of an organism. In highly dynamic situations the CRF may happen to operate, the very infrastructures of these systems may need to be dynamic too, as a function of the external environment and system's mission, which may change over time. We may

---

Ruling Distributed Dynamic Worlds, by Peter S. Sapaty
ISBN 0-471-65575-9 Copyright © 2005 John Wiley & Sons, Inc.

need not only to update the infrastructure or recover lost parts but also (regularly or accidentally) create a completely new one, as a reaction on changing situations and conditions, and this should be carried out without interrupting the work of the system and loss of its integrity.

WAVE-WP, with its dynamic, fully interpreted and integral nature, absolute, viruslike mobility in computer networks, and possibility of creating spatial "holographic" infrastructures and algorithms not connected in advance to any physical resources, may be a challenging candidate for the organization and management of advanced CRF.

In this chapter, we will consider the run time creation and modification of different kinds of distributed command-and-control infrastructures for mobile dynamic systems, with orientation on crisis management, and also solving some basic control and management problems in a spatial mode, using and navigating these infrastructures in parallel.

7.1 DISTRIBUTED CREATION AND RECONFIGURATION OF AN INFRASTRUCTURE

Imagine we have a distributed set of units, say, crisis reaction ones of any nature, as shown in Figure 7.1, assuming that each one is supplied with a copy of WAVE-WP interpreter and these interpreters can communicate.

7.1.1 Hierarchical Infrastructure

Also imagine we need to set over these units a certain hierarchical infrastructure, reflected, say, by oriented links named infra, with unit.1 chosen as top of the hierarchy, as shown in Figure 7.2. Through this infrastructure, the units may be able to pass

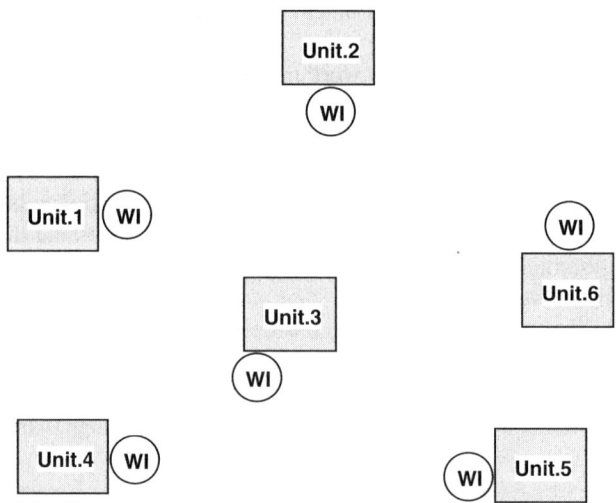

Figure 7.1. Distributed set of units with WAVE-WP language interpreters.

CREATION AND RECONFIGURATION

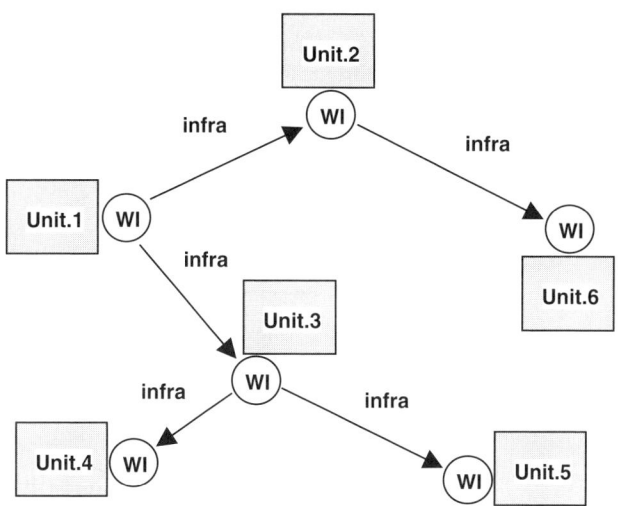

Figure 7.2. Hierarchical command and control infrastructure to be built.

commands and data to each other, control command execution by subordinate units, report the progress or failure to superior units, assess and generalize distributed situations with making proper decisions at each level, and so on.

Any infrastructure topology can be represented in WAVE-WP syntax in a compact manner, as a graph template, and created in parallel and distributed mode by spreading this template in the space of units. The following program, based on one of many possible templates for the infrastructure of Figure 7.2 and applied from any unit, creates the needed infrastructure in a space-covering mode:

```
direct # unit.1;
create(
  (+infra # unit.2; +infra # unit.6),
  (+infra # unit.3; +infra # (unit.4, unit.5))
)
```

Parallel and stepwise evolution of this program within the set of units in a creative template-flow mode is shown in Figure 7.3 (where worked parts of the program are being omitted, and rule `create` is inherited during the template's splitting, modification, and propagation between the units).

7.1.2 Other Topologies: Centralized, Chain, and Ring

Any other infrastructure topologies, both imaginable and so far unimaginable, can be created by self-evolving WAVE-WP patterns. Some examples are presented below.

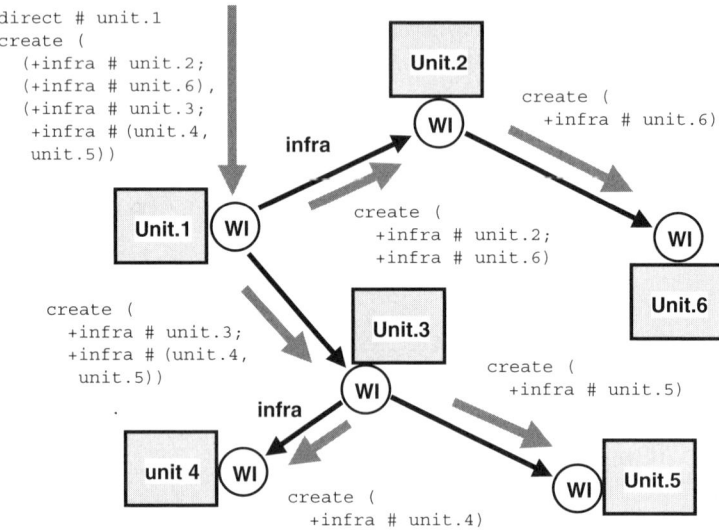

Figure 7.3. Run time creation of the infrastructure in a template-flow mode.

Centralized. A centralized infrastructure, with one unit (say, unit.3) as a central node, and all others directly subordinate to it, can be set up by the following program:

```
direct # unit.3;
create(+infra # (unit.1, unit.2, unit.4, unit.5, unit.6))
```

The parallel navigation process and the infrastructure created by this program are shown in Figure 7.4.

The program can be simplified by not naming subordinate units explicitly if, say, all these units are lying within certain vicinity (let it be 1000 m) from the central unit, as follows:

```
direct # unit.3; create(+infra ## r1000)
```

If the units in Figure 7.4 are the only ones accountable and accessible in the system, the program may be written as:

```
direct # unit.3; create(+infra ## any)
```

If any unit can be a central one in the infrastructure, and the program can be applied exactly from this unit (unit.3 included), it can be further simplified as follows:

```
create(+infra ## any)
```

Chain. Oriented chain, as a degenerated hierarchy, where each unit (except the first and the last ones) has exactly one superior node and one inferior node, can be

CREATION AND RECONFIGURATION

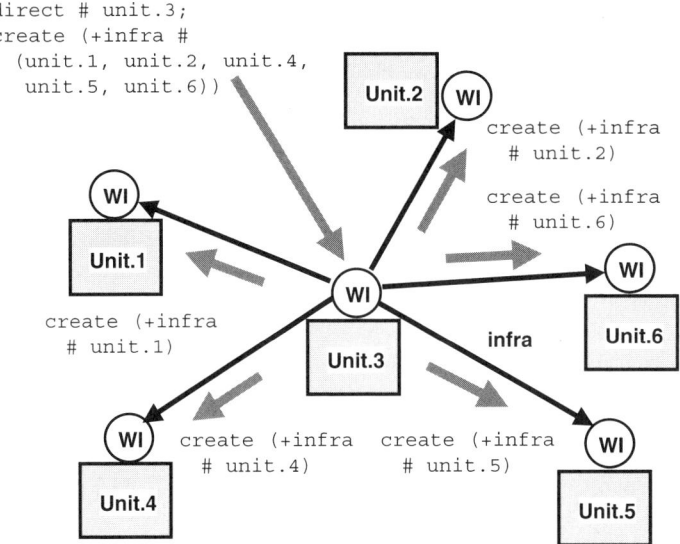

Figure 7.4. Centralized infrastructure.

created by representing its graph template explicitly, similar to the hierarchy shown earlier, as follows for the possible chain depicted in Figure 7.5:

```
direct # unit.1;
create(
  (+infra ## unit.4; +infra ## unit.5; +infra ## unit.3;
  +infra ## unit.6; +infra ## unit.2
)
```

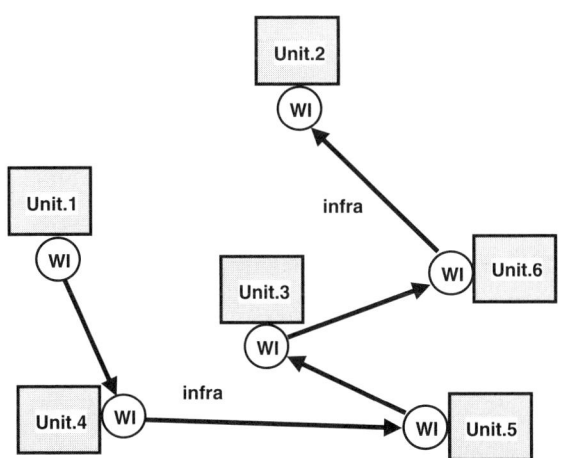

Figure 7.5. Chain infrastructure.

As the same link name is used for the whole infrastructure, a cyclic program for the chain creation may be more elegant, using a vector of ordered node names, and not repeating explicitly the link name in the text for each step, as follows:

```
direct # unit.1;
Fn = (unit.4; unit.5; unit.3; unit.6; unit.2);
repeat(Fn != nil; create(+infra # Fn : 1); Fn : 1 = nil)
```

It will terminate when the vector `Fn` with a sequence of node names becomes empty, after regularly removing its first element used for hops between units. Such a program can also be much shorter than the previous one if the number of units becomes large enough.

To form an arbitrary chain starting from any unit and passing all other units, where any unused unit may happen to be the next one, the following program (repeatedly choosing the first unit replying of no links with other units, unless there are no such units) will suffice:

```
repeat(
  orparallel(direct # any; (any # any)== nil);
  [create(-infra # BACK)]
);
```

Ring. An oriented ring infrastructure can be used for different service purposes, also as a supplement to the traditional hierarchical infrastructures. The ring infrastructure differs from the chain in that there should be a link from the last node in the chain to the first one, and the chain-building programs above can be easily extended to add such a link. For example, the latest program forming an arbitrary chain can be modified for the oriented ring as follows:

```
repeat(
  orparallel(direct # any; (any # any) == nil);
  [create(-infra # BACK)]
);
create(+infra # INFRA_start)
```

A possible infrastructure produced by this program is shown in Figure 7.6 (assuming, for simplicity, that the same units have been selected at each step as by the cyclic chain-building program discussed before).

7.1.3 Infrastructure Modification

Any created infrastructure can be modified in WAVE-WP by another template (or a number of templates operating in parallel), which can evolve on the already existing infrastructure topology. This modification may reflect, say, a changed subordination and communication within the crisis reaction force while adjusting to the new conditions, goals, and environment.

For example, for the infrastructure in Figure 7.2, to start from `unit.5` and remove the existing link to `unit.3` in parallel with the creation of a new oriented `infra` link

CREATION AND RECONFIGURATION

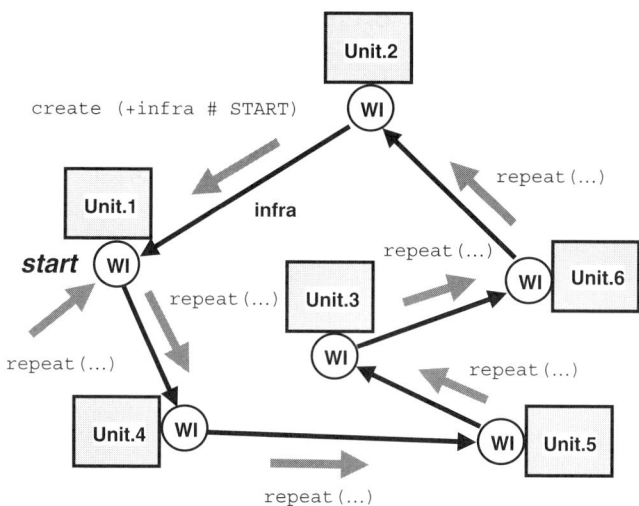

Figure 7.6. Ring infrastructure.

from unit.2 to unit.5, the following program will be sufficient, with its work depicted in Figure 7.7:

```
direct # unit.5;
(any # unit.3; LINK = nil), create(-infra # unit.2)
```

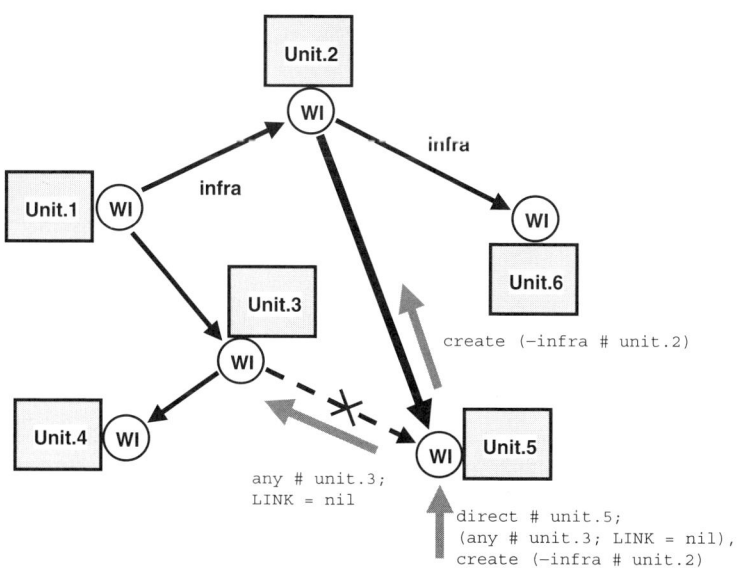

Figure 7.7. Run time reconfiguration of the infrastructure.

7.2 DYNAMIC HIERARCHY BASED ON PHYSICAL NEIGHBORHOOD

Dynamic infrastructures for advanced crises reaction forces, especially robotized ones, with some examples shown in the previous section, can be based on a variety of principles. Let us take here into account the peculiarities of the physical distribution of the set of units in space (each having physical coordinates) as shown in Figure 7.8, where each unit box, for simplicity, is numbered rather than named, and also assumed containing the WAVE-WP interpreter.

Let us consider the run time creation of a hierarchical infrastructure based on the neighborhood between units, where the top of the hierarchy should be assigned to a geographically most central unit, with subordinate units lying in a certain physical vicinity of their direct superior units. Such an infrastructure may have the most rational organization in dynamic situations, where higher levels of hierarchy are better protected from an unwanted outside influence, with top unit being the safest one.

7.2.1 Finding the Most Central Unit

We start with finding the topologically most central unit, in a parallel and distributed manner. Applied from any unit and entering in parallel all units (the starting one included, i.e., reentering it), the following program finds the network address of the most central unit, assigning it to nodal variable Nstart in the application unit, as shown in Figure 7.9.

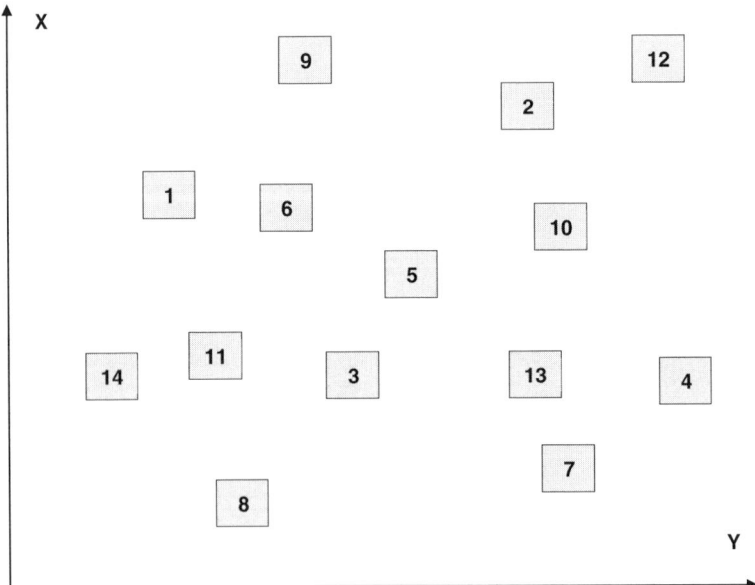

Figure 7.8. Set of units with coordinates in the physical world.

HIERARCHY BASED ON PHYSICAL NEIGHBORHOOD

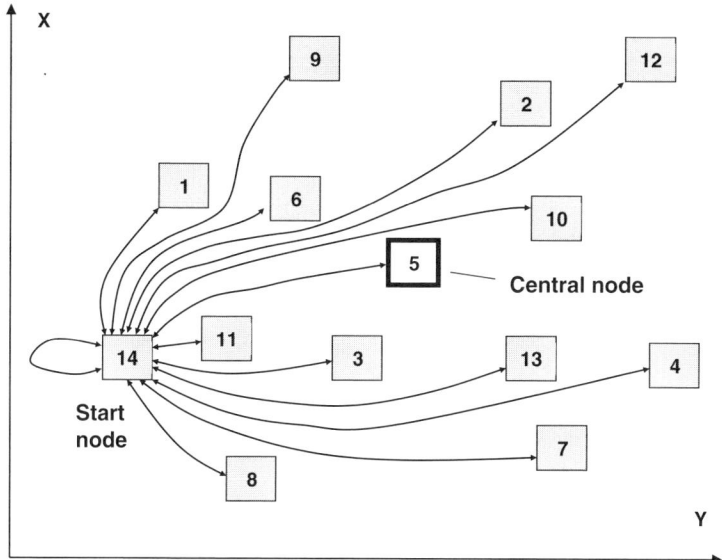

Figure 7.9. Parallel finding of the central unit.

```
Faver = average(direct # all; WHERE);
Nstart =
min(direct # all; (Faver, WHERE) ? distance_ADDRESS): 2
```

The program uses the echo rule `average` to find the average values among x and y coordinates of all units and also external procedure `distance` to determine deviation of each unit from this averaged x-y position, the latter coordinates brought to the units in frontal variable `Faver`. The unit with the smallest deviation is considered the most central one. Echo rule `min` is used for this parallel and distributed process, where the addresses of units are aggregated with the deviation values (the latter compared with each other by rule `min` in a distributed mode).

7.2.2 Creating Infrastructure from the Center

In the next step, from this central unit (with its address in `Nstart` in the unit where the previous program was applied), a new hierarchical infrastructure can be built with direct subordinate units assumed lying within a given range from the superior units (say, 40 m). Asynchronous and parallel breadth-first space navigation mode underlies the following program for doing this job:

```
direct # Nstart; Frange = r40; N = 1;
repeat(
  direct ## Frange; grasp(N == nil; N = 1);
  [create(-infra # BACK)]
)
```

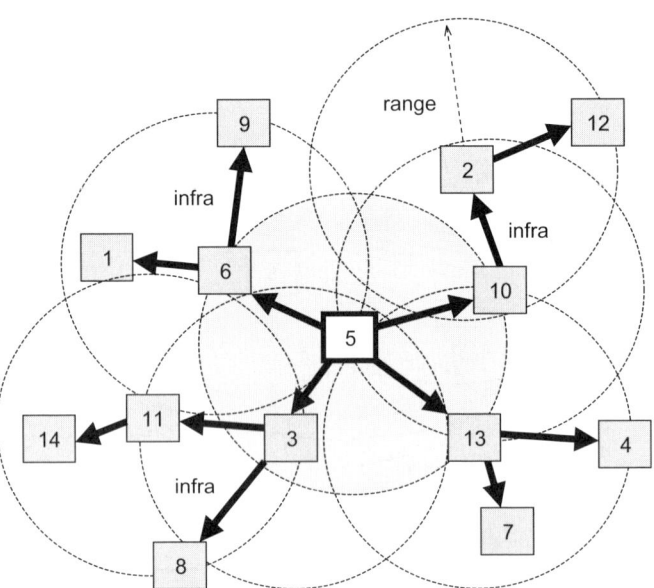

Figure 7.10. Creating a neighborhood hierarchical infrastructure.

The rule grasp seizes the nodal resources and keeps them solely until the code it embraces terminates (the latter checking the node mark kept in nodal variable N, and marking the node if it is not marked). This resolves a possible conflict between different branches entering the same units when creating the next level of the hierarchy. A possible resultant infrastructure built by this program is shown in Figure 7.10.

This infrastructure may not be optimal as, for example, it contains chains of units 5–10–2–12 and 3–11–14 (rather than preferred hierarchies). Changing the physical range within which the directly subordinate units should reside (possibly, adjusting it individually for different nodes), also introducing more intelligent link-building strategies into the program above (which can be effectively done in WAVE-WP) we can receive more suitable infrastructure solutions, say, as the one shown in Figure 7.11.

The two programs discussed above, that is, finding the central unit and creating neighborhood infrastructure from it, can work together as one program, being initially applied from any unit. We can also put the united program in a global cycle (with a period, say, 360 sec), each time updating the infrastructure—actually creating it again from a possibly new central unit. As the units can potentially be on a move, changing both absolute and relative positions, this will constantly keep the infrastructure updated and optimized.

The resultant program, starting from any unit and doing all this, also deleting the previous infrastructure links before creating the new ones, will be as follows:

```
repeat(
  Faver = average(direct # all; WHERE);
  Nstart = min(direct # all;
```

BASIC COMMAND-AND-CONTROL SCENARIO

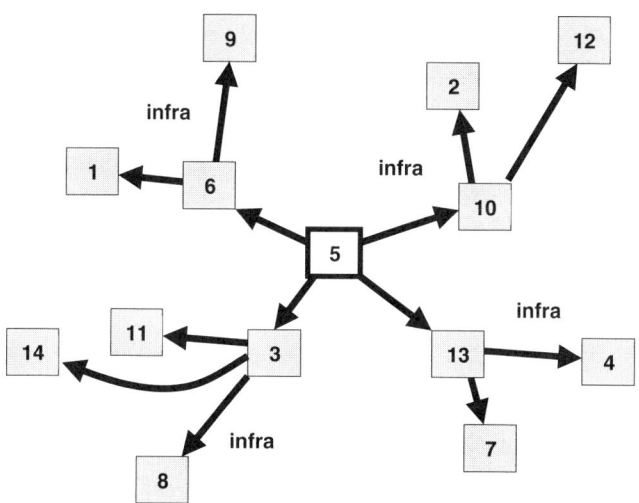

Figure 7.11. Possible optimization of the neighborhood infrastructure.

```
  (Faver, WHERE) ? distance_ADDRESS) : 2;
direct # Nstart; Frange = r40;
quit(
  N = 1;
  repeat(
    direct # Frange; grasp(N == nil; N = 1);
    [any # any; LINK = nil];
    [create(-'infra' # BACK)]
  )
); USER = ADDRESS; TIME += 360
)
```

The rule quit allows us to clean up the distributed system from unwanted temporary information (here node marks represented by variable N, as well as internal tracks), preparing it for a next spatial iteration on updating the infrastructure. The program also sends outside the system, to the symbolic user (represented by environmental variable USER), the network address of the new central node, as a new top of the hierarchy.

7.3 BASIC COMMAND-AND-CONTROL SCENARIO IN WAVE-WP

7.3.1 Recursive Hierarchical Command and Control (CC)

Any command-and-control systems, distributed ones included, can be created, managed, and simulated in WAVE-WP. The following program, using the hierarchical infrastructure discussed and created above (see Fig. 7.2), implements a classical CC process, where global command or mission scenario, applied to the top of the hierarchy, is analyzed and executed stepwise in a top-down manner, with proper actions taking place at each level. Acknowledgments to the executed commands are moving bottom-up, being analyzed at higher levels.

At each level, the commands are executed according to the peculiarities of this level and then transformed and modified for the level below, being subsequently replicated and sent to all direct subordinates for further execution and modification. Only after full completion of the command on its level and getting acknowledgments from all units of the subordinate level (the latter units doing the same for their own subordinates, and so on, due to recursion), is a unit allowed to report to its direct superior about the completion of its command.

7.3.2 Implementing CC in WAVE-WP

The following program, implementing an hierarchical command-and-control process, is based on a recursive navigation procedure FcommandControl, which, during its spatial evolution, activates at each level three specific external procedures.

The first procedure, command, is activated in a top-down spread of the program, executing the command in frontal variable Fcommand descending from the layer above, while taking into account peculiarities of the current level (the growing level value being registered in another frontal variable Flevel). The second procedure, transform, modifies and details the current level command for it to be appropriate for the subordinate level. The third one, control, is activated at each level only after all subordinate levels (to their full depth, due to spatial recursion) complete the execution of the commands delivered to them, making final control arrangements at this level.

```
FcommandControl = {
  Flevel += 1;
  sequence(
    ((Fcommand, Flevel) ? command;
      Fcommand = (Fcommand, Flevel) ? transform;
      +infra # any; ^ FcommandControl
    ),
    Flevel ? control
  )
}
```

The procedures command, transform, and control may have fully human execution, human participation and interaction, or can be fully automatic.

The activation program launching recursive procedure FcommandControl, applied from any unit and then starting the distributed CC process from unit.1, may be as follows:

```
Fcommand = 'top command or mission scenario';
direct # unit.1; ^ FcommandControl
```

The distributed hierarchical command-and-control process, set up by this program for the previously built hierarchy shown in Figure 7.2, is illustrated in Figure 7.12.

BASIC COMMAND-AND-CONTROL SCENARIO

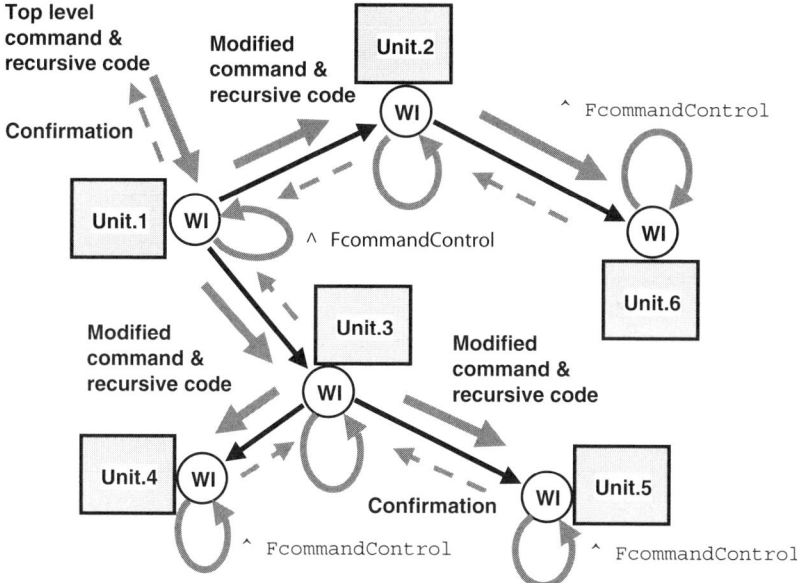

Figure 7.12. Hierarchical command-and-control scenario in WAVE-WP.

7.3.3 Adding Payload to the CC Implementation

Any "payload" to this recursive distributed command-and-control scheme can be effectively added in WAVE-WP. For example, we can modify the previous program that it should collect all failures of the execution of commands at each level, along with all its subordinate levels, exhibiting these separately to the outside user at each level, after the completion of command-and-control process at it. The following program uses nodal variables Nlocal and Nunder in each unit to collect command failures, respectively, in the current unit and in all subordinate units:

```
FcommandControl = {
  Flevel += 1;
  sequence(
    (Nlocal = (Fcommand, Flevel) ? command;
      Fcommand = (Fcommand, Flevel) ? transform;
      +infra # any; ^FcommandControl
    ),
    (USER = Nlocal; USER = Nunder;
      (BACK #; Nunder) &= Nlocal & Nunder
    )
  )
}
```

The command-and-control processes described above can also start from the most central node of the hierarchical neighborhood infrastructure built previously (see Fig. 7.10), using the node's network address regularly supplied by the spatial infrastructure update process.

7.4 SOLVING DISTRIBUTED MANAGEMENT PROBLEMS

7.4.1 Hierarchical Resource Management

Using the automatically created hierarchical infrastructures, let us consider the solution of another very traditional problem: Supply and resupply of some physical resource to different units of the distributed force via these infrastructures (which may, e.g., be food, gasoline, or munitions). The resource may be limited and therefore should be divided between units according to the local needs, with physical delivery to proper units, say, from a storage associated with the top unit, and then down the hierarchy. The following program executes this resource management scenario via the hierarchical infrastructure:

```
Fexplore = {
  [Namount = Nresource ? amount;
   Nrequest = Needed - Namount;
   +infra # any; ^Fexplore];
  (BACK #; Nrequest) += Nrequest
};
direct # unit1;
repeat(
  [^Fexplore; Nrequest != 0;
    repeat(
      Fsupply =
        or((Namount > Nrequest; Nrequest), Namount);
      +infra # any; Nrequest != 0;
      Fwithdraw =
        Fsupply * Nrequest/(BACK #; Nrequest);
      Nresource +=
        (BACK #; (Fwithdraw, Nresource) ? withdraw)
    )];
  TIME += 360
)
```

It makes regular top-down checks of the demands in units, as a difference between the needed and actual levels of the resource, using recursive spatial procedure Fexplore. The program sums up and returns these demands back through the hierarchy, analyzing on the top level the difference between the resource level in the storage and the received sum of demands.

The resource, as a physical matter, is kept in nodal variable Nresource at each unit, and external procedure amount determines the level of this resource (put into another variable Namount holding information only). The nodal variable Needed holds the level of the resource needed for each unit for its normal work, and using the value in Namount it is possible to define how much of the resource is to be requested for the unit (the latter kept in nodal variable Nrequest at each unit). The ultimate value in Nrequest in each unit, after the termination of recursive procedure Fexplore, is a sum of individual request if the unit and the requests of all units subordinate to it.

Another top-down parallel process by this program makes decisions about the amount of the resource to be forwarded to each unit, physically delivering this resource. If there is enough resource received by the upper unit, the needs are satisfied in full. Otherwise the allowed amount for a unit depends on the amount of the resource in the upper unit, demands from peer units on the same level, sum of demands from the level subordinate to this unit, and the unit's own needs.

The calculated value of the resource to be withdrawn from the upper unit, in favor of the current unit, is stored in frontal variable Fwithdraw, with procedure withdraw executing the physical operation. The withdrawn resource is added to variable Nresource holding physical matter in the current unit.

The parallel top-down and bottom-up operation of this program over the hierarchical infrastructure of Figure 7.2 is symbolically shown in Figure 7.13, with program code, information, and physical matter flowing along the created infrastructure links.

7.4.2 More Complex Management Scenarios

In the previous section, we considered only a single-source, parallel, and distributed management scenario working on the hierarchical infrastructure. Much more complex, multisource, and cooperative solutions of complex management problems arising in the distributed CRF may be effectively organized in WAVE-WP.

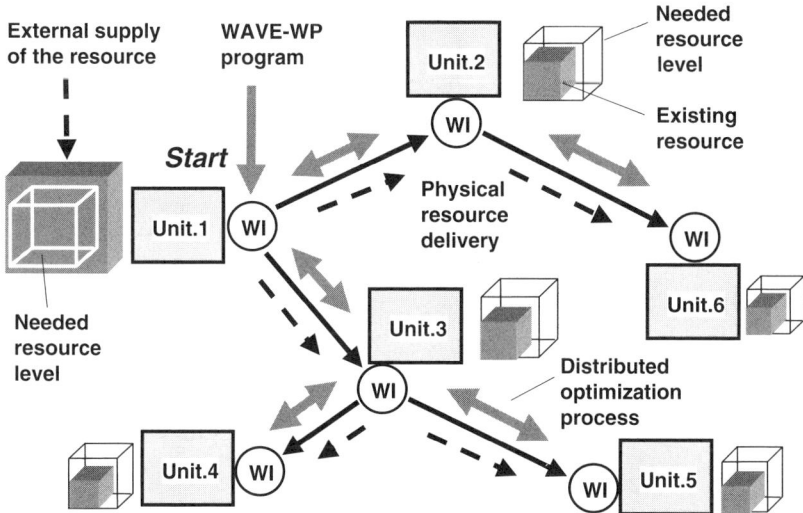

Figure 7.13. Physical resource management and delivery via the infrastructure.

A sketch is shown in Figure 7.14 where the initiative starts independently from four different points, with four separate optimization scenarios simultaneously evolving via the same infrastructure. Imagine these scenarios must find a satisfactory solution for all units that launched them (i.e., unit.1, unit.4, unit.5, and unit.6) by negotiations via the infrastructure, spreading their own operations and data to other units at run time.

Scenarios 2, 3, and 4 may, for example, work on local problems in the units that started them and local vision of their solutions, whereas scenario 1, launched on the top level, may moderate the solutions on other scenarios, in an attempt to find the overall optimum. This optimum may take into account the results of local optimizations and may need relaunching of local optimization processes within the global balancing process. All these processes may be highly interactive, with involvement of human personnel.

As can be seen from Figure 7.14, the locally issued scenarios may invoke a nonlocal optimization as, for example, scenarios 3 and 4 via the superior for them unit.3. Scenario 2, starting at unit.6, spreads activity one way only, to unit.2, and launches an optimization process for unit.6 there, whereas the final solution is found and brought back to unit.6 only by the global scenario 1, which regularly navigates the hierarchy in both ways and also coordinates the interaction between scenarios 3 and 4. Efficient program code in WAVE-WP can be written for these and many similar cases, adding any needed semantics to this spatial dynamics and evolution.

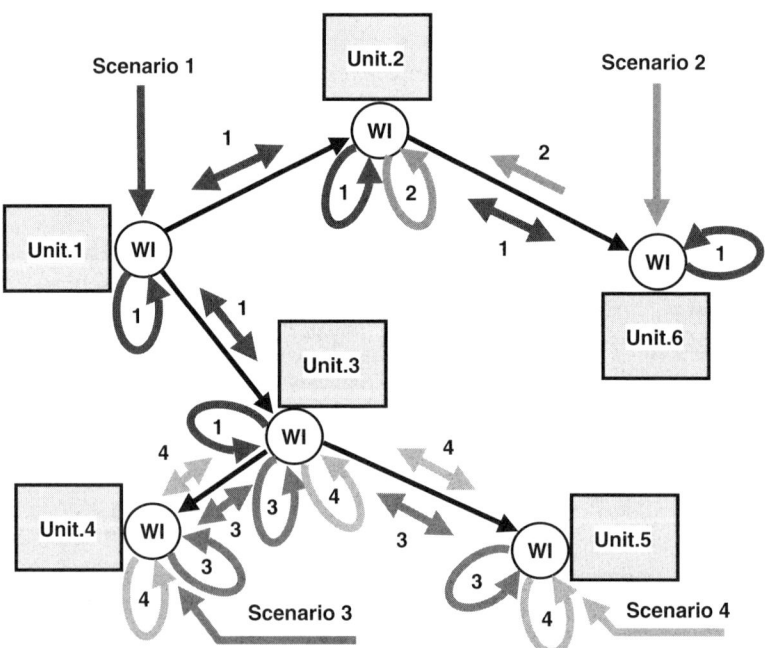

Figure 7.14. Cooperative management with multiple interacting scenarios.

7.5 AIR TRAFFIC MANAGEMENT IN DYNAMIC ENVIRONMENTS

One of the strictest applications of WAVE-WP may be air traffic management in crisis situations, where a priori traffic schedules may become useless, and radar stations and control centers may be indiscriminately damaged. Quick run time recovery and reintegration into the functional whole may be required, with dynamic routing and collision avoidance performed by smart local, rather than traditional global, mechanisms. We will show here how WAVE-WP can be used for the traffic management and control in dynamic situations, on an example of tracking aerial objects in distributed spaces by mobile intelligence.

Usually air space is divided into regions inside which traffic management is provided by local control centers with their own radar sets, with human controllers working cooperatively to provide smooth traffic flow in the distributed airspace. Neighboring centers and radar units may communicate and hand over aerial objects when the latter cross regional boundaries.

7.5.1 Creation of the Radar Neighborhood Infrastructure

Radar stations and control centers usually do not cover the whole aerial region (say, in the scale of a country) alone and must communicate with each other to keep the overall observation integral and continuous. In case of indiscriminate damages, also physical movement of mobile radars, it will be very useful to provide run time establishment and regular update of their neighborhood infrastructure. The virtual links between units in such an infrastructure should reflect the fact that their radar stations cover adjoining or overlapping regions of space.

We will show here a program that puts a process into each unit that regularly checks physical distance to other units, and if it is lower than the sum of their radar ranges (with effective range, say, 40 km), a nonoriented `neighbor` link should be set up between the units. On the other hand, if the `neighbor` link already exists, but physical distance between the nodes exceeds the sum of their ranges (i.e., the radar nodes have moved apart in the meantime), such a link should be removed.

The following program, initially put into and then working in parallel with all units, dynamically creates and regularly updates the radar neighborhood infrastructure (with the chosen period of 360 sec):

```
direct # any;
repeat(
Flocation = WHERE;
  [direct # any; ADDRESS < BACK;
    or(
      ((Flocation, WHERE) ? distance<=80000;
        or(neighbor # BACK, create(neighbor # BACK))
      ),
      (neighbor # BACK; LINK = nil)
    )];
  TIME += 360
)
```

In this program, the creation of new links and removal of the outdated ones is allowed by only one of the adjacent nodes—to prevent unwanted competition between them. An example of the created and updated neighborhood infrastructure is shown in Figure 7.15.

Using WAVE-WP model, such regularly updated neighborhood infrastructure can essentially help us formulate and simplify the solutions of different discovery, tracking, analysis, and hand over tasks in a fully distributed network mode, without any central resources.

7.5.2 Mobile Tracking of an Aerial Object

Fully mobile and parallel character of WAVE-WP allows each aerial object to be discovered and accompanied with an individual mobile intelligence propagating via the computer network space, whereas the objects chased are moving in physical space, as shown in Figure 7.16.

The spatial tracking process regularly checks visibility of the object from the current radar station. If the object (say, aircraft) disappears the neighboring stations search for it, using the created neighborhood infrastructure. And if such a station is found, the whole tracking process, with accumulated knowledge about the object, moves into this station and continues observing the object from there. A basic WAVE-WP code for this distributed tracking process may be as simple as follows:

```
Fobject = TRW562;
repeat(
  repeat(Fobject ? seen; TIME += 10);
  any # any; Fobject ? seen
)
```

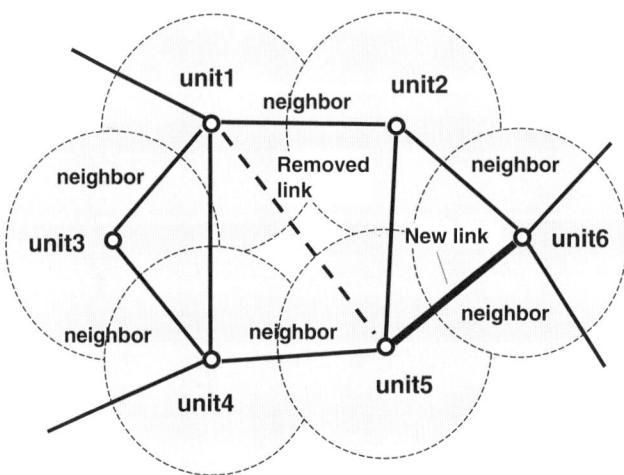

Figure 7.15. Dynamically updated radar neighborhood infrastructure.

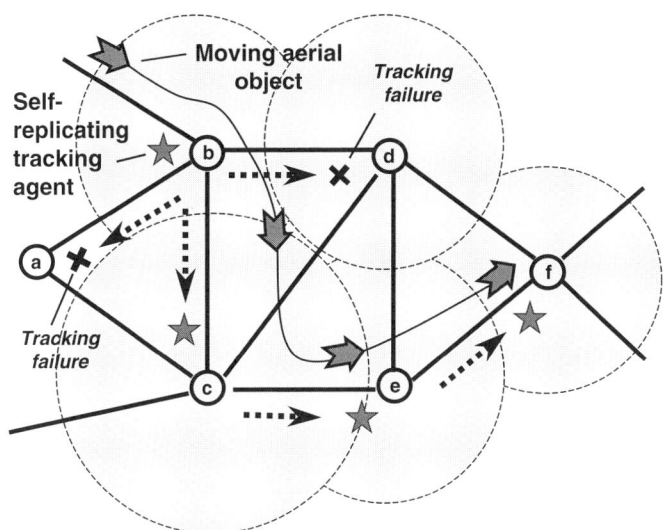

Figure 7.16. Distributed networked tracking of aerial objects.

In this program, it is assumed that the controlled object has identity TRW562, its presence is checked by the current radar station every 10 sec (using external procedure seen, resulting in failure if the object cannot be observed by the radar), and the object's identity is carried throughout the network in frontal variable Fobject. During this distributed tracking, the mobile intelligence may accumulate and carry with it any payload needed (fully representing the individual aircraft, for its distributed chase and management, including the management strategy and tactics).

7.5.3 Simultaneous Multiple Tracking

Many objects can be observed and analyzed simultaneously in such a way, as shown in Figure 7.17. The individual tracking processes can spatially communicate with other such processes, as well as with global coordination mechanisms, which can be mobile too. All this can provide effective air traffic control in fully dynamic environments, without a priori connection to particular management centers (which is very vital as the latter can be indiscriminately damaged in crises situations).

The previous program can be easily updated for dealing with multiple objects (actually all possible objects seen) after their discovery, while starting in all network nodes (radar stations) in parallel, as follows:

```
direct # all; ? alienObjects;
Fobject = VALUE;
repeat(
  repeat(Fobject ? seen; TIME += 10);
  any # any; Fobject ? seen
)
```

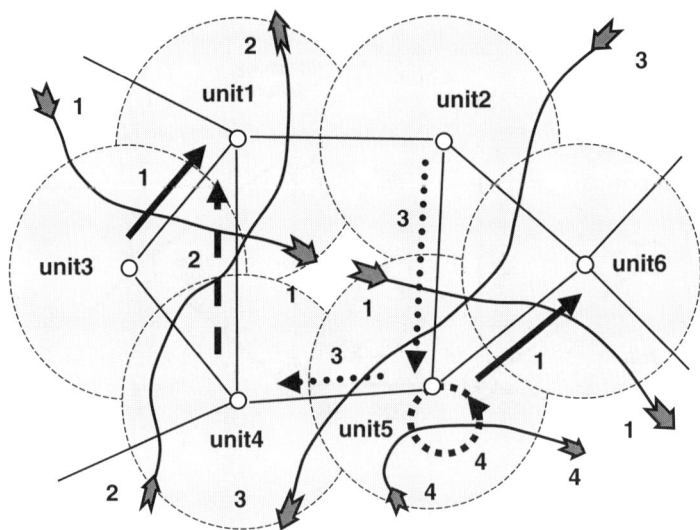

Figure 7.17. Simultaneous tracking of multiple objects.

The external procedure `alienObjects` results in failure if no objects are found in a node, otherwise producing a vector with elements being identities of the objects discovered. For each such object (each scalar *value* of the vector) the rest of the program will be replicated, with the object's personal identity kept in individual frontal variable `Fobject`, similar to the previous program for tracking a single object.

7.5.4 Setting Up Global Control

The above program is replicated for each node and each object discovered, with the created copies working fully independently while migrating in the network space. If needed, we can set up any nonlocal analysis and control over these distributed and self-evolving mobile processes in WAVE-WP. For example, we may want to know when all these processes terminate and report this to a global operator, and this can be easily accomplished by the following extension of the previous program:

```
sequence(
  (direct # all; ? alienObjects;
   Fobject = VALUE;
   repeat(
     repeat(Fobject ? seen; TIME += 10);
     any # any; Fobject ? seen
   )
  ),
  USER = 'object tracking terminated'
)
```

AIR TRAFFIC MANAGEMENT

Or we may want to have the ability to terminate all these distributed mobile tracking processes at any time, for which the following program will be sufficient:

```
orparallel(
  (direct # all; ? alienObjects; Fobject = VALUE;
  repeat(
    repeat(Fobject ? seen; TIME += 10);
    any # any; Fobject ? seen
    )
  ),
  repeat(Nstop == nil)
)
```

The program uses a competitive "terminator" process (a loop organized by `repeat` rule), being the second branch of the `orparallel` rule. Cancellation of the whole multiple tracking mission can be easily done by terminating this second branch, say, by entering the starting node with:

```
Nstop = 1
```

After this, the second branch of `orparallel` will immediately report itself as a successful one (as `repeat` always terminates with success, due to the language semantics). This immediately aborts the first branch, which represents a multiple tracking algorithm.

7.5.5 Other Traffic Management Tasks

The WAVE-WP technology can effectively provide many other useful services for the distributed air traffic management and control, which may include but are not limited to:

- Distributed assessment and collective awareness of the existing local or global aerial situations, with making intelligent distributed automated or fully automatic decisions.
- Distributed interactive simulation of the air traffic in any region, and especially in emergency situations.
- Outlining and sealing dangerous regions (say, caused by air pollution, thunderstorms, or wars).
- Distributed dynamic finding of optimal routes for aircraft, emergency rerouting, collision avoidance, and the like.
- Tracking and analyzing particular air vehicles by mobile intelligence, with certain actions to be taken if needed, say, their destruction, and many others.
- The mobile techniques described can also be efficiently applied for tracking and chasing different kinds of objects in the networks of highways, railways, and waterways, as well in distributed computer networks, with run time solutions of complex routing and traffic and topology optimization problems.

7.6 CONCLUSIONS

We have considered run time creation and modification of different kinds of infrastructures of distributed dynamic systems, allowing us to organize them internally, and also provide efficient external supervision, command, and control.

Using peculiarities of the WAVE-WP model and language, these infrastructures can be made dynamic, adaptable to the rapid changes in the environment, internal organization of the system, and external goals. They can be easily updated and recovered after the loss of system components, as well as completely rebuilt from scratch at run time, without interruption of the system functionality and integrity.

A number of command-and-control and other system management tasks over these dynamic infrastructures, including those effectively solved by recursive navigation of arbitrary hierarchies and self-propagating and replicating mobile intelligence in computer networks, have been analyzed and described in WAVE-WP, operating in a parallel and fully distributed mode.

Due to full mobility of WAVE-WP code on the implementation layer, *any* distributed infrastructure and *any* distributed system can be created, updated, and recovered starting from *any* of its working component, which is particularly important if parts of the system can be indiscriminately damaged, say, in hostile environments.

8

MORE CRISIS MANAGEMENT SCENARIOS AND SYSTEMS

This chapter, as sequel to the previous one describing management problems and their solutions in changing environments, concentrates on a variety of dynamic and specific situations and scenarios that may take place in multirobot, computer network and large united global systems—with the relation to advanced crisis management. All these are investigated and expressed using the WAVE-WP model and language.

These scenarios include single, two-, and multiple-robot patrol of regions, where robots can cooperate in crisis situations; multirobot hospital service with both scheduled and emergency scenarios, where different types of robots can work together; future combat systems widely using robotic components and having network-centric organization; and top-level computer network management with finding–collecting infected and congested links and nodes and an attempt to infer the virus sources.

Possible distributed national and international defense and security infrastructures integrating local crisis management centers, computer networks, radar stations, mobile unmanned vehicles and systems, and the like are outlined too. Higher level, cognitive systems are also discussed, with their expression on a semantic level and dynamic implementation in parallel and fully distributed mode by multiple robots.

Ruling Distributed Dynamic Worlds, by Peter S. Sapaty
ISBN 0-471-65575-9 Copyright © 2005 John Wiley & Sons, Inc.

8.1 REGION PATROL BY MOBILE ROBOTS

A typical example of using both single and multiple robots with orientation on crisis situations may be the patrolling of a certain region, where human operators (or some superior control systems) are being alarmed in case of the discovery of an irregular situation (say, an "alien" object has been detected by infrared sensors mounted on robots). We may suppose, for simplicity, that robots follow prescribed routes given by a sequence of waypoints represented by physical coordinates.

8.1.1 Patrolling by a Single Robot

Patrolling a region by a single robot cyclically, continuously following the given sequence of waypoints (where after the last point the first one is used again), is shown in Figure 8.1.

The WAVE-WP program cyclically organizing single-robot movement between the waypoints (where any robot, say, Robot.1 can be used for this) will be as follows:

```
Npoints = (x4_y3, x7_y3, x9_y7, x5_y10, x2_y6);
repeat(
  WHERE = Npoints : 1;
  Npoints &= Npoints : 1; Npoints : 1 = nil;
  [? infraredCheck; USER = 'alarm'_WHERE];
)
```

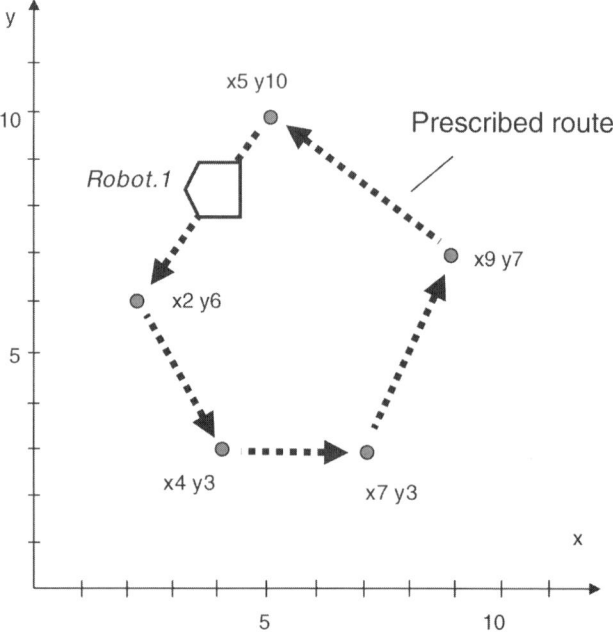

Figure 8.1. Patrolling of a region by a single robot.

In this program, : is an indexing act, and & is the act that appends a scalar or a vector to the end of another vector, and an assignment of nil to an element of the vector removes this element. These operations keep the vector in Npoints as an endless cyclic sequence, where repeated assigning of its rotating first element to the environmental variable WHERE always forces the robot to move to this new physical location.

The specific external procedure infraredCheck (with resultant state thru if it discovers an intrusion, allowing subsequent moves to take place by the language semantics), which can be written in any other language, is called from the wave program by act ?. USER is another environmental variable of WAVE-WP, enabling us to send information outside the system (here as an aggregate value consisting of the word alarm and current PW coordinates of the robot). And the square bracketed part of the program (same as using rule none with parentheses) always leaves the resultant program control in the same node regardless of the wave it represents.

8.1.2 Simultaneous Region Patrol by Two Robots

We can easily modify the previous example in the way that (an extended) region should be simultaneously patrolled by two robots, each one cyclically following its own route, given by the sequence of waypoints too. Such a situation is shown in Figure 8.2, where Robot.1 and Robot.2 are used for the parallel patrol.

Each robot should report to the user or an external system (which may be the same or different for different robots) if its sensors discover unusual objects, as for the single

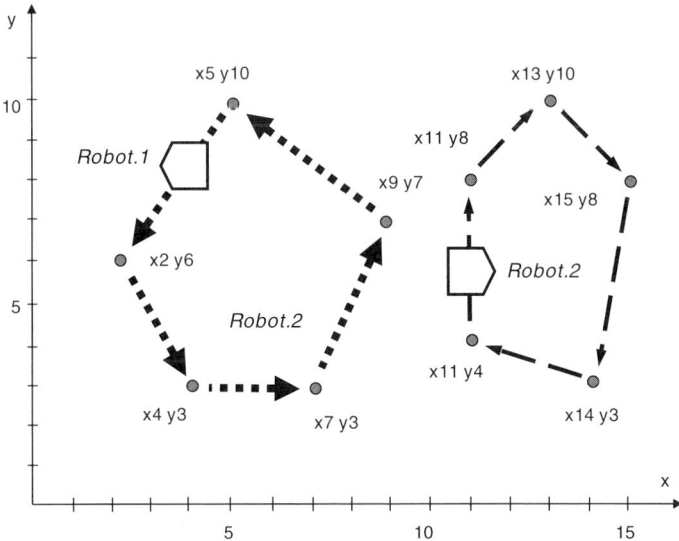

Figure 8.2. Patrolling of an extended region by two robots.

robot before. The WAVE-WP program for these two particular robots will be as follows:

```
(direct # robot.1;
  Npoints = (x4_y3, x7_y3, x9_y7, x5_y10, x2_y6)),
(direct # robot.2;
  Npoints = (x11_y4, x11_y8, x13_y10, x15_y8, x14_y3));
repeat(
  WHERE = Npoints : 1;
  Npoints &= Npoints : 1; Npoints : 1 = nil;
  [? infraredCheck; USER = 'alarm'_WHERE];
)
```

8.1.3 Possible Cooperation Between the Two Robots

We can also organize different forms of cooperation between the robots. For example, a robot discovering irregularities first may contact another robot and change the recorded waypoints in the latter to its own waypoints. This will make the two robots operate together on the same route, thus strengthening the intrusion-checking capabilities on the route where intrusion has already been detected.

We may also allow robots to follow other robot's waypoints for some time only, after which they should return to their native routes (until a new emergency situation occurs in themselves or in the other robot, with requesting the other robot's help or being requested again, respectively).

The above program for the two robots can be modified for such a cooperative work as follows:

```
(direct # robot.1;
  Npoints = (x4_y3, x7_y3, x9_y7, x5_y10, x2_y6)),
(direct # robot.2;
  Npoints = (x11_y4, x11_y8, x13_y10, x15_y8, x14_y3));
Np = Npoints;
repeat(
  WHERE = Np : 1; Np &= Np : 1; Np : 1 = nil;
  free(
    ? infraredCheck; Fp = Np; USER = 'alarm'_WHERE;
    direct # any; Np = Fp;
    TIME += 1800; Np = Npoints
  )
)
```

In this program, when the crisis situation occurs, its branch embraced by the rule `free`, while entering the other robot, substitutes its waypoints stored in nodal variable `Np` by the new ones brought in the frontal variable `Fp` (the latter carrying copied waypoints of the requesting robot). These after 30 min (or 1800 sec) will be restored to the native waypoints, previously saved in the robot in nodal variable `Npoints`.

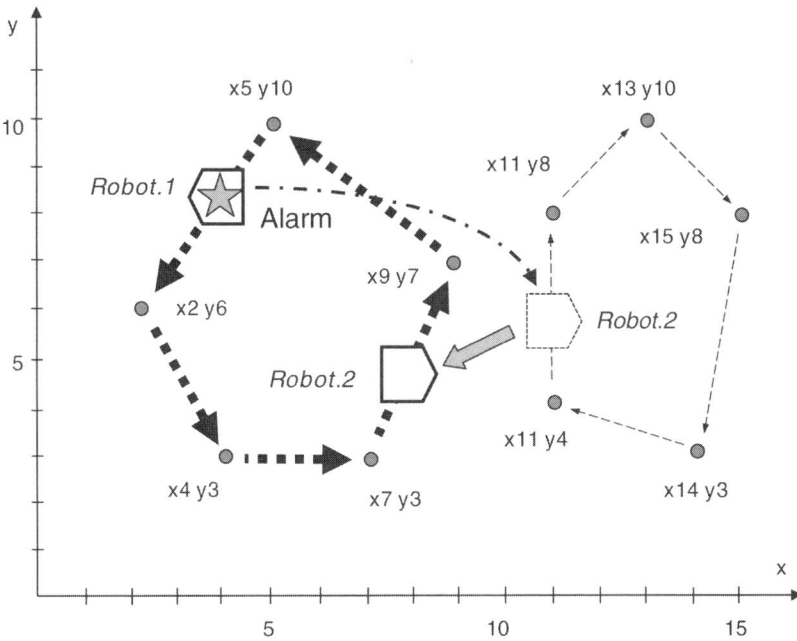

Figure 8.3. Changing the route of Robot.2 after Robot.1 discovers intrusion.

The situation managed by this program, where Robot.2 changes its route to the one of Robot.1 after the latter discovers an intrusion, is shown in Figure 8.3. After the "leave" time expires, Robot.2 returns to its own route and continues its endless, cyclic patrol there.

8.1.4 Dynamic Patrol by Any Number of Robots

In the previous program, we loaded two robots, Robot.1 and Robot.2 explicitly, and the program could work only if these particular robots were available and used. For highly dynamic and crises-prone situations, with limitations on the open and changeable hardware, and with its possible indiscriminate damages, a much more flexible strategy may be needed to be chosen. For example, the patrol scenario may first look for available robots by itself (with its different parallel branches, possibly, competing for a mobile hardware to be seized), and then self-loads into them, subsequently using these robots for the patrol.

Such a strategy may effectively shift the main mission emphasis from the robots to the mission scenario, where *the scenario should survive by any means and fulfill the mission objectives*, rather than pieces of mobile hardware that can be destroyed!

The scenario can also be generalized to be suitable for any number of waypoint sequences to be followed simultaneously (and, correspondingly, any number of mobile robots used, when available). We will also assume that when a robot follows its waypoint sequence and finds an intrusion, it tries to engage for its own route additionally only one of other busy robots (the latter should not be in the alarm state and not already engaged by other robots), possibly, competing in this with other similar crisis-discovering robots. If this attempt fails, the robot should continue operating on its route alone, as usual.

The following extension of the previous program will satisfy all these new options and demands:

```
Fpoints = (x4_y3, x7_y3, x9_y7, x5_y10, x2_y6),
Fpoints = (x11_y4, x11_y8, x13_y10, x15_y8, x14_y3),
... ... ... ... ... ... ... ....;
or(direct # all; grasp(Nbusy == nil; Nbusy = 1));
(Np, Npoints) = Fpoints;
repeat(
  Nalarm = nil;
  WHERE = Np : 1; Np &= Np : 1; Np : 1 = nil;
  free(
    ? infraredCheck; Nalarm = 1;
    USER = 'alarm'_WHERE; Fp = Np;
    or(
      direct # any;
      grasp(
        Nbusy == 1; Nalarm ==  nil;
        Nrequested == nil; Nrequested = BACK
      )
    );
    Np = Fp; TIME += 1800;
    Np = Npoints; Nrequested = nil
  )
)
```

In this program, for sequentially checking all existing robots (including the one in which the program starts) of being not busy—for seizing one of them (using special constant `all` in hops and state variable `Nbusy`), or all other busy robots not in the alarm state and not engaged by alarmed robots (with `any` in hops and state variables `Nalarm` and `Nrequested`), the sequential rule `or` is used. This rule stops the search with the first successful branch (the braches forming dynamically), thus moving into the first robot found. The rule `grasp` protects common nodal resources during the competitive access to them from parallel branches.

We can easily improve this program further, say, for bringing another robot to the current robot's route, which is physically closest to the current robot, also for requesting any needed number of such additional robots (including all others not in the alarm state and not used by other alarmed robots). The program also can be easily modified that in case of the absence of suitable robots, it can indefinitely repeat the search for them, as these robots may eventually appear in the open system, and so on.

A variety of other, much more complex and advanced forms of cooperation between unmanned patrol units in crisis situations can be effectively programmed in WAVE-WP.

8.2 DISTRIBUTED DYNAMIC COGNITIVE SYSTEMS

8.2.1 Semantic Representation of Distributed Cognition

Cognitive systems belong to the most advanced class of intelligent systems—the ones aware of what they are doing. While cognitive systems include reactive and deliberative processes, they also incorporate mechanisms for self-reflection and adaptive self-modification. Traditional cognitive systems usually orient on single robots, both ideologically and by original implementation, and making dynamic and intelligent groups of them, especially operating in crisis situations, can meet serious difficulties. This may also need a cardinal revision of the basic principles of cognitive systems themselves.

The WAVE-WP model and technology allow us to describe reactive, deliberative, and reflective processes of cognitive systems on a much higher, semantic level, often revealing all potential parallelism and distributed character of the whole cognitive mission (rather than system), effectively abstracting from any possible implementation. All these processes can be set up directly over (and in) the distributed environment in an integral and seamless form, as shown in Figure 8.4.

Using this approach, the general mission of a cognitive system can be described and constantly maintained as a highly organized whole, with automatic implementation effectively delegated to the WAVE-WP interpreter, the latter providing dynamic mapping of this whole onto robotic hardware. The technology can involve at run time any needed (or available) number of robots to fulfill the mission's objectives, with collective behavior of robots often emerging as a mere derivative of the mission parallel and distributed interpretation process.

The mission-to-hardware mapping process may be fully distributed and automatic, not requiring central resources. Failed robots can be automatically substituted at run time without the loss of overall mission integrity. In Figure 8.5, a possible snapshot of migrating spatial cognitive processes is shown over the group of four robots.

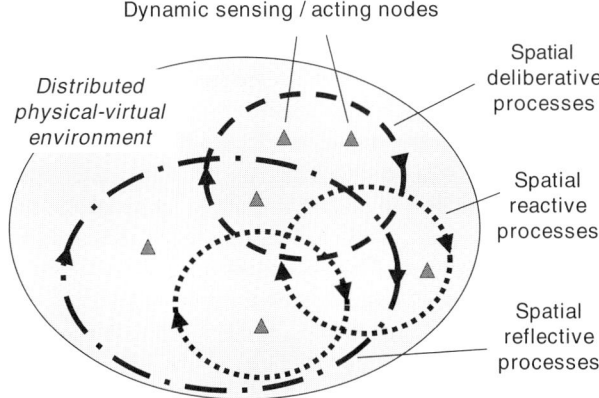

Figure 8.4. Fully distributed, semantic, representation of cognitive processes.

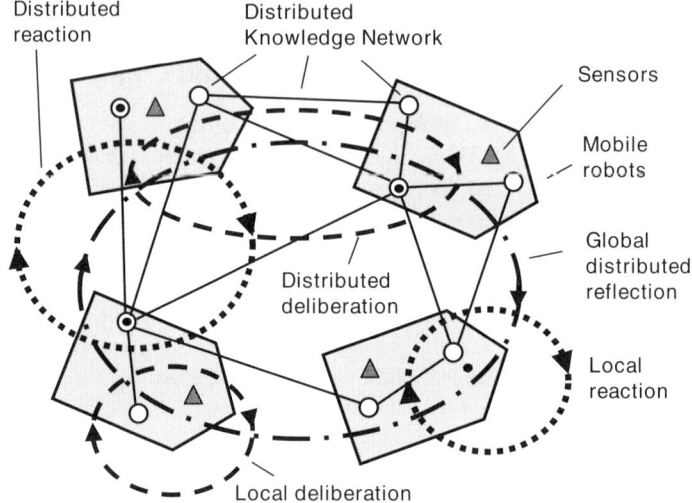

Figure 8.5. Distributed dynamic mapping onto the world of robots.

Some of the processes may happen to be local to robots, while others covering groups of them or the whole community; and this distribution and localization of processes in and between the robotic bodies may vary at run time.

8.2.2 Multirobot Patrol as a Distributed Cognitive System

In the previous section, we described a distributed region-patrolling system supported by multiple cooperative robots. It may have some features of a distributed cognitive system, comprising both distributed *deliberative* processes, where multiple robots perform intrusion checking on prescribed routes, and also distributed reactive processes where the robots that found the intrusion reprogram other robots to follow their own routes, for a certain period of time.

The behavior of these dynamically reprogrammed robots is definitely a *reactive* one, where the robots that discovered the intrusions within their deliberative processes are behaving as *distributed sensors* for the reprogrammed robots, as well as for the whole system.

We can also easily set up any higher level, or *reflective* processes, analyzing the work of the whole distributed system, due to the fully interpretative nature of WAVE-WP and unlimited access to other, including running, processes. As a simplest example, the following program, working independently and in parallel of all other processes of the distributed patrolling system, collects (as aggregate values) the names of robots involved in this process, their current coordinates, current time, and current alarm state (i.e., whether the robot is processing the intrusion scenario at this moment in time).

```
USER = (direct # all; DOER_WHERE_TIME_Nalarm)
```

The other elementary program may collect (as aggregate values too) the names of robots that have discovered and are currently processing the intrusions, along with the names of all other robots that have been diverted in an emergency to the current robot's route. These values may also be followed by this common route itself (in its current state as an endless sequence). This top-level analysis can be made regular, cyclic, with certain interval, say, 60 sec, as follows:

```
repeat(
  USER =
    (direct # all; Nalarm != nil;
    DOER & (direct # any ; Nrequested == BACK; DOER) &
    Npoints % '_'
  );
  TIME += 60
)
```

Any other and, of course, much more sophisticated and intelligent reflective processes expressing "self-consciousness" or "meta-system" level of such distributed robotic systems can be implemented in WAVE-WP, which can themselves have a highly parallel and fully distributed nature.

8.3 MULTIROBOT HOSPITAL SCENARIOS

8.3.1 A Robotized Hospital

A hypothetical robotized hospital is depicted in Figure 8.6, where different types of mobile robots such as cleaning (C), life support (L), and body state checking (S) ones may serve patients simultaneously and cooperatively. The robots can autonomously move between patients avoiding obstacles and each other, accessing a store of drugs and a litter box.

Both scheduled and emergency scenarios are possible for such a hospital in WAVE-WP. State-checking robots may regularly visit patients and measure their body temperature, blood pressure, and heart beat. Cleaning robots may periodically search the territory, collecting litter (waste) and discharging it into the litter box. Life support robots may deliver prescribed medicine picked up from the store to patients on a schedule or an emergency request. Emergency scenarios may originate from patients asking for unscheduled assistance or from robots themselves (say, the state-checking ones) discovering nonstandard situations.

Scheduled and emergency scenarios engaging different robots simultaneously may result in their cooperative behavior, with some simple patterns reflected by arrows in Figure 8.6. Life support or state-checking robots may, for example, call cleaning robots to collect accidental litter or waste from patients. Life support robots may be called by both state-checking robots and patients, as well as by other life support robots. More than one robot may serve the same patient at the same moment in time. Alarms may also be issued to human medical staff in complex cases.

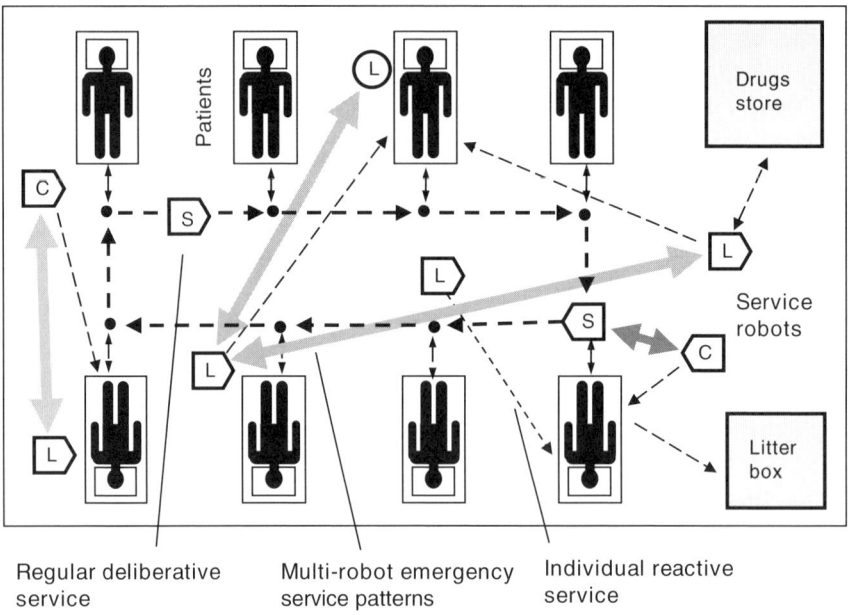

Figure 8.6. Multirobot hospital service.

8.3.2 Hospital World Representation

The formalized structured physical–virtual–execution (or PVE) world of the robotized hospital described in the previous section is shown in Figure 8.7. The route for movement of state-checking robots with access to beds of all patients is represented as a directed loop of physical–virtual (or PV) nodes, having names (just as bed numbers) and physical coordinates, these nodes being interconnected by oriented links named next.

The drugstore and litter box (as drugs and litter, for short) are PV nodes too, with both names and physical coordinates. Robots are represented as execution (or E) nodes, with specific names and arbitrary initial distribution in physical space given by coordinates too (not shown in the figure).

This combined PVE world can be described and created by the following parallel program:

```
(Fstart, Fbed) = 1;
Froute = (x1y3; x2y3; x3y3, x4y3, x4y1, x3y1, x2y1, x1y1);
create(
    direct # (
    drugs_x5y4, litter_x5y0,
    robot.1_x4.5y1.5, robot.2_x3.5y3.2, robot.3_x3.7y2.4,
    robot.4_x4.7y2.6, robot.5_x3.5y1.2, robot.6_x3y1.8
    ),
```

MULTIROBOT HOSPITAL SCENARIOS

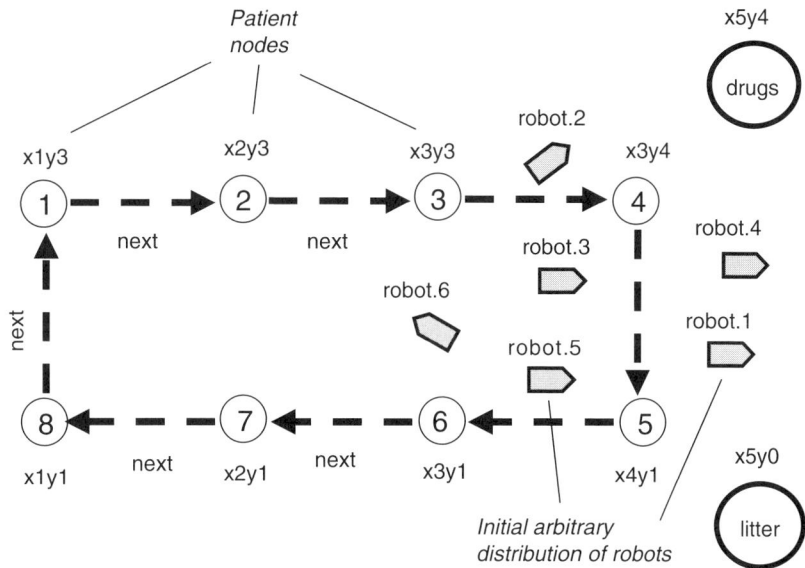

Figure 8.7. Formalization of the hospital world.

```
(direct # Fbed_(Froute : 1);
repeat(
   (Froute : 1) = nil; Froute != nil;
   Fbed += 1; + next # Fbed_(Froute : 1)
);
+ next ## Fstart
)
)
```

In this program, drugs, litter, and all robotic nodes are created and distributed in parallel. Simultaneously with this parallel process, the loop of connected PV nodes representing beds of patients is created stepwise, with the final node linked to the start one.

Representation of this PVE world may be kept in a centralized form, say, in a separate host computer or in the brain of one of robots engaged; it can also be seamlessly distributed between any number of robotic brains and used as a common distributed knowledge processed in parallel in WAVE-WP.

8.3.3 State-Checking Scenario

Let us consider implementation of the regular state-checking scenario to be executed by state-checking robots. The scenario is assumed to start simultaneously in the bed-related nodes 1 and 5, and then its two independent branches should independently, asynchronously, and endlessly move along the same cyclic route, always staying in its different points (waiting in the current place if the next place of the route is occupied). When staying in bed nodes, they check the states of patients using specific (external to WAVE-WP) procedure `stateCheck`.

The WAVE-WP program for this scenario will be as follows, where `robot.5` and `robot.2` are recommended to be used as the system state-checking resources for the scenario execution (otherwise the interpretation system may be trying to use default resources if available):

```
RESOURCES = (robot.5, robot.2); Fstart = (1, 5);
direct # Fstart;
repeat(
  Noccupied = 1; ? stateCheck;
  wait(+ next # any; Noccupied == nil);
  (BACK #; Noccupied) = nil
)
```

8.3.4 Cleaning Scenario

For the hospital floor-cleaning scenario, we will assume that starting from some arbitrary point (with coordinates in `Fstart`) a random choice is regularly made (using external function `randomChange`) of shifts in x and y coordinates—as `dx` and `dy` in the language syntax. These coordinate changes should be done in such a way that the new positions lie in a certain range (given in `Frange`) from the previous positions, to perform the cleaning process smoothly.

In each new position, the floor should be cleaned within the position's vicinity (its range given in `Fvicinity`), using external procedure `cleanPickup`. The latter also picks up a possible waste (litter) at this location, adding it to the already collected and carried waste. If the waste accumulated is essential (say, exceeding 2 kg), the scenario should interrupt and move directly to the litter box (or `litter`, in our formalization), discharging it there (using external procedure `release`) and returning subsequently to the point it left.

Making this scenario multiple by starting simultaneously from two physical world points with certain coordinates, and recommending the interpretation system to use `robot.1` and `robot.6` for the cleaning purposes, the WAVE-WP program will be as follows:

```
RESOURCES = (robot.1, robot.6);
Fstart = (x1.5y0, x4.5y1.5);
Frange = 0.5; Fvicinity = 0.25;
direct # Fstart;
repeat(
  Fwaste += (Fvicinity ? cleanPickup);
  [Fwaste > 2.0; direct # litter; Fwaste ? release];
  direct # (Frange ? randomChange)
)
```

Fwaste, carrying physical matter, will be automatically emptied by releasing into the litter box, so after returning to the cleaning loop it will have zero value.

8.3.5 Life Support Scenario

Life support scenario may be a reactive one, where a specialized service should be performed as a response to an unscheduled event (say, dangerous condition discovered by a state-checking robot or an emergency call by the patient herself).

In this request, a detailed patient investigation should be made (using, e.g., advanced electronic or chemical equipment), by which a trip to the drugstore (or drugs) may be needed for picking a proper medicine, subsequently applying it to the patient (like giving pills or making injection). In more complex situations, other robots may be called for the assistance, or human personnel alarmed for performing more sophisticated actions or making nonlocal decisions.

In the simplest form, such a life support scenario, applied in one the patient nodes (i.e., 1 to 8) with an unusual situation, and recommending robot.3 or robot.4 to be used for the support functions, may look as follows:

```
RESOURCES = (robot.3, robot.4);
Nresult = (? detailedCheck);
  (Nresult : 1 == otherRobotCall; ......),
  (Nresult : 1 == humanCall; ......),
  (Nresult : 1 == applyMedicine; Fmedicine = Nresult : 2;
    (direct # drugs; Fmedicine ? pickUp) ? apply
  )
```

The result of the detailed check is put into the variable Nresult as a vector of two elements, with the first element identifying a subsequent action to be performed and the second one detailing what is needed for this action. For the third option of the first element, that is, applyMedicine, the medicine name to be taken from the drugstore (drugs) is assigned to frontal variable Fmedicine. After moving to drugs, the corresponding physical matter is picked up, brought back to the request point, and applied to the patient.

8.3.6 Multirobot Service Snapshot

A snapshot of activity and possible distribution of the recommended robots (i.e., robot.1 to robot.6) during execution of the above-mentioned scenarios is depicted in Figure 8.8. The state-checking robots 2 and 5 are currently serving patients in beds with numbers, respectively, 6 and 2, and life support robots 3 and 4 are serving patients in beds 6 and 4 where, for example, robot.3 is called by state-checking robot.2, and robot.4 is working on a direct request from the patient. The casual positions of the cleaning robots, performing jobs within the given vicinity of randomly selected physical nodes, are shown in the picture too.

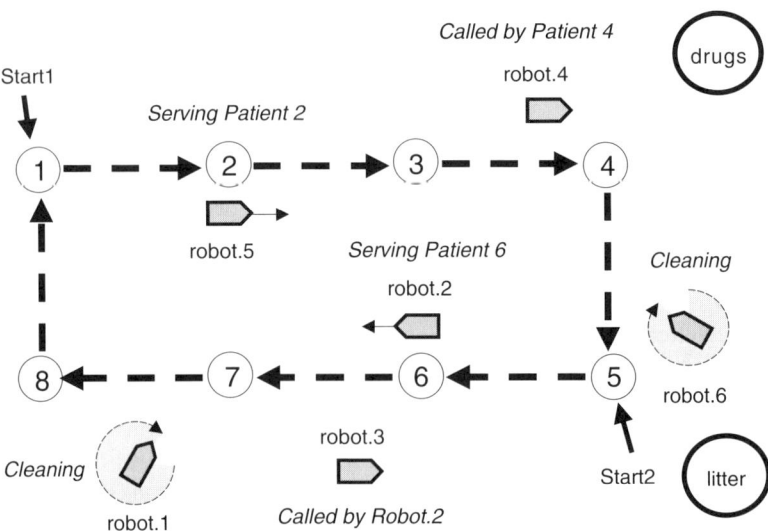

Figure 8.8. Possible snapshot of the multirobot hospital service.

Much more complex scenarios involving multiple robots of the same or different orientation, working together with each other and with the human personnel, can be described and implemented in WAVE-WP in a parallel and distributed manner. Emergency scenarios may range from local ones, serving individual patients, to global campaigns such as, for example, evacuation of the whole hospital due to natural or artificially caused disasters.

8.4 FUTURE COMBAT SYSTEMS

Another, quite different, application of WAVE-WP is shown in Figure 8.9, with installing communicating copies of the WAVE-WP interpreter in main units of the future combat systems (FCS). These may extensively use multiple sensor and execution robots, with a network-centric global organization.

8.4.1 Advantages of Using WAVE-WP

Due to the high robotization and tendency to drastically reduce direct participation of the human personnel on battlefields (using it mostly for higher level decisions and global control), with possible exclusion of humans from the operational theater at all, FCS may be effectively used in extremely dangerous and dynamic situations, with high probability of the loss of components, which cannot be afforded when using manned forces.

FUTURE COMBAT SYSTEMS

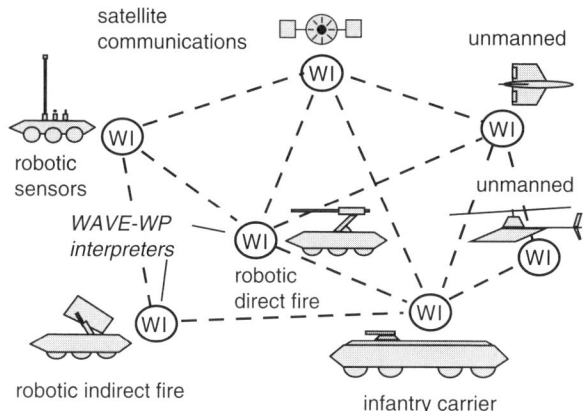

Figure 8.9. Future combat systems integrated in WAVE-WP.

System infrastructures in such applications can be partially or even completely destroyed, and their recovery without interrupting the mission may be of paramount importance. Examples of such dynamic infrastructures were shown in the previous chapter in WAVE-WP, where any unit could happen to appear at any level and at any moment in time. This could be unthinkable in case of the manned armies, where a general might happen to be subordinate to a soldier!

The evolution of FCS into fully unmanned armies in the future will need radically different military doctrines and different ideology of command and control (CC), with much more dynamic and intelligent CC infrastructures—undoubtedly more advanced than the traditional hierarchies. Such infrastructures can be modeled and investigated in WAVE-WP, and in the previous chapter we showed other possible organizations and their easiness to be expressed in the spatial language.

For the FCS, coexisting multiple dynamic infrastructures may prove useful within the same system, with switching between them as a reaction on changing environments and goals, and possible run time transformation and evolution of one infrastructure into the other, and so on.

WAVE-WP can be used as a top ideological and technological basis for the integration, management, and control of FCS, providing the highest possible integrity and self-recoverability of such forces as a system, making them flexible and easily task oriented. The technology can also protect FCS from unwanted external influences, such as, for example, aerial attacks, using networked mobile radar stations as part of the advanced forces.

8.4.2 Target Fusion and Distribution by the Infrastructure

Let us consider, as a very simple example of a possible work of FCS as one system, a regular fusion of information about the enemy targets coming from multiple distributed sensors, with subsequent distribution of this collected data to all firing units and their autonomous selection of appropriate targets to fire.

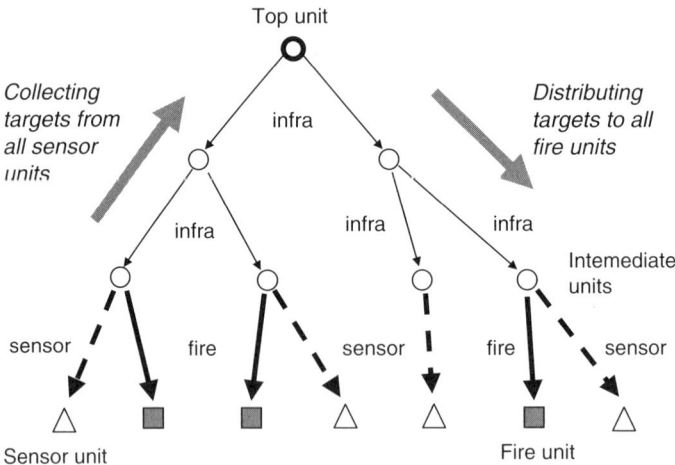

Figure 8.10. Bottom-up collection of targets with their top-down distribution for firing.

We assume that the internal CC infrastructure of FCS has a hierarchical nature, having a top control unit (or `topUnit`) providing global CC of the whole mission, intermediate units enabling local CC on different levels, and also fringe units comprising (unmanned) sensor and fire units. The top and intermediate units are interconnected by oriented links named `infra`, and the fringe sensor and fire units are connected to the former units by oriented links named, respectively, `sensor` and `fire`, as shown in Figure 8.10.

The top unit may correspond, for example, to a manned infantry carrier, and the intermediate units may be both manned and unmanned units, the latter including unmanned aerial and ground vehicles (UAVs and UGVs) with automatic CC capabilities. At the lowest level, the special mobile robotic sensors and direct and indirect fire robots can be used. Such an infrastructure may be highly dynamic, adjusting to the mission's evolution and changeable parameters of the hostile environment; it can be effectively built and maintained in WAVE-WP at run time, as discussed in the previous chapter.

Starting from all sensor units that see the potential targets, the target data is moving upwards the hierarchy, being merged in intermediate units, as well as the top one. Upon completion of this bottom-up process, the target data collected in the top unit is sent down the hierarchy, replicating in intermediate nodes, unless reaching the fire units. The latter may choose the most appropriate (or reachable in principle) targets themselves and can process them independently from each other, possibly in parallel, where the same targets may happen to be shot from different fire units.

8.4.3 Fusion–Distribution Scenario in WAVE-WP

The parallel top-down access of all distributed sensors, parallel bottom-up collection of all visible targets from these sensors, subsequent (also parallel) distribution of the collected target information to all distributed fire units, and firing targets from the latter can be easily provided by the following simple program in WAVE-WP. This program works with an arbitrary number of levels of the CC hierarchy and its structure.

```
direct # topUnit;
repeat(
  Fseen = (
    repeat(+infra # any); +sensor # any;
    ? detectObjects
  );
  Fseen != nil;
  [repeat(+infra # any); +fire # any;
    Fseen ? selectShoot]
)
```

The external procedure `detectObjects`, working in a sensor unit, leaves as a resultant value a vector of discovered targets. These are collected from all sensors by an internal echo process triggered by the assignment to variable `Fseen` in `topUnit`. The external procedure `selectShoot` analyzes targets delivered to each fire unit in the frontal variable `Fseen`, choosing the most appropriate ones to shoot. Top-down movement through the whole CC hierarchy is easily organized by the parallel spatial loop (using `repeat` rule) through directed `infra` links, finally using `sensor` or `fire` links.

The parallel bottom-up movement and target merge are performed automatically by the distributed WAVE-WP interpreter, via internal tracks left after the top-down process. Using the global spatial loop in this program (by the outer `repeat`), the distributed target collection with distributed firing can continue indefinitely, until all targets are destroyed or become invisible.

In a more advanced version of this scenario, the intermediate units of the hierarchy and the top one can actively participate in a more intelligent targets collection, analysis, and their distribution to the most appropriate fire units.

Any other types of infrastructures, such as, say, the ring one described in the previous chapter, can also be used for the collection of targets from multiple sensors and their distribution to multiple fire robots in WAVE-WP.

8.5 CRISES MANAGEMENT IN OPEN NETWORKS

8.5.1 Embedding Distributed WAVE-WP System

The WAVE-WP technology can be incorporated in open networks by installing multiple copies of the language interpreter in its most important and sensitive points, as depicted in Figure 8.11. The interpreters can communicate with each other, forming in and between them task-oriented runtime knowledge and command and control infrastructures, which can be effectively used for both local and global management of distributed computer networks.

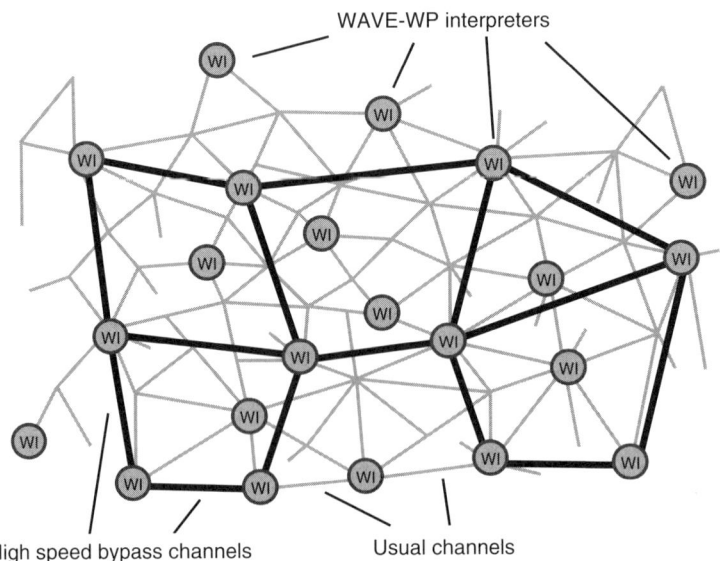

Figure 8.11. Installing WAVE-WP interpreters and high-speed bypass channels.

A network of these interpreters, as already explained throughout this book, forms a highly parallel and distributed machine spatially executing high-level WAVE-WP language. Additional high-speed bypass channels may be set up between the interpreters, as a sort of meta-network overlaid on top of the existing computer network, as shown in Figure 8.11. WAVE-WP interpreters can communicate via both usual and bypass channels, with the former predominantly used for checking links and node accessibility as well as traffic measurement, and the latter—in irregular and emergency situations—also for top-level and safe network control.

8.5.2 Establishing Higher Management Layer

WAVE-WP can efficiently integrate traditional network management tools and systems, as shown in Figure 8.12, dynamically extracting higher level knowledge from raw data via them. Establishing a higher, intelligent layer, allows us to analyze varying network topologies, regulate network load, and redirect traffic in case of line failures or congestions.

Any other, local or distributed systems and network management tools can be integrated in WAVE-WP and cooperatively used in a coordinated manner, as one system. The integrated local facilities may serve as multiple sensors and effectors of the distributed WAVE-WP system.

The spatial model can provide automatic awareness and assessment of distributed situations, making autonomous decisions on higher levels, especially in crisis situations. A distributed crisis management scenario can start on an initiative of WAVE-WP

CRISES MANAGEMENT IN OPEN NETWORKS

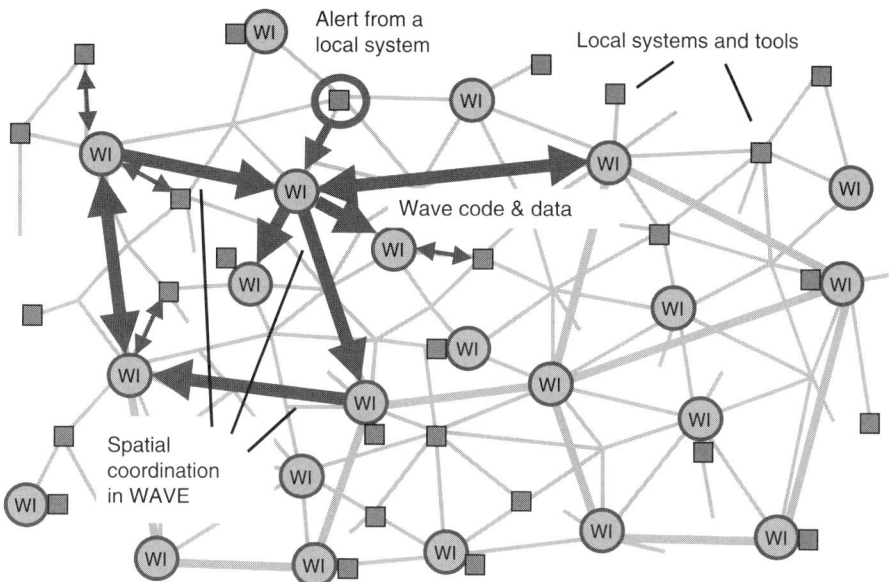

Figure 8.12. Integration with local management systems and tools.

system within the regular network supervision process, or it can be triggered by local specialized systems serving as emergency sensors, as shown in Figure 8.12.

WAVE-WP can automatically outline and cut off infected regions, inferring attack sources using large volumes of knowledge about virus attacks in parallel. It can make detailed postmortem analysis of attacks, forming and accumulating distributed logical attack patterns. The technology can integrate existing antivirus tools into one distributed system with a unified control and management. Let us consider some elementary code examples in relation to these problems.

8.5.3 Collecting All Infected Nodes

The simplest solution for this case may be by using direct and parallel access to all network nodes allowed by WAVE-WP, lifting their network addresses if some special external procedure virusCheck discovers that the computer has viruses, and subsequently collecting all such addresses in one point (as a vector), as follows:

```
USER = (direct # all; ? virusCheck; ADDRESS)
```

It is supposed that virusCheck results in control state thru if the computer node is infected, thus allowing a subsequent move to take place, which is just the node address (provided by environmental variable ADDRESS), to be returned and collected.

The solution above shifts all organization of parallel and distributed network access and data collection exclusively to the WAVE-WP interpreter (to its default internal mechanisms). We may also want to put more details to the application layer and choose the strategy that might be best for the case, writing, say, a spanning tree coverage of the whole network and collecting data from nodes accessed in such a way. We may also want to collect together with the node addresses the detected virus names, the latter represented as values of virusCheck, as follows:

```
direct # start;
USER =
  repeat(
    N == nil; N = 1;
    (? virusCheck; VALUE_ADDRESS; ! done), any # any
  )
```

8.5.4 Finding Congested Links and Nodes

Let us consider finding all links in the network passing that exceed some threshold time, and also all nodes that respond to an access of them via all associated links with a delay exceeding a threshold too. Such nodes with high probability may themselves be congested and represent a source of delay, rather than links leading to them. Node congestion may also be an indication that these computers may be infected by viruses. An example of congested parts of the network is shown in Figure 8.13.

The following program, using breadth-first spanning tree coverage of all network links (rather than all nodes, as in the previous example), collects link names (along with the node addresses on both sides) if passing through them exceeds the threshold (say, 300 sec). The program also counts in all nodes the minimum delay of response to all neighboring nodes, that is, after passing all links leading to the nodes.

After the spanning tree coverage completes, the accumulated minimum response time in each node is checked of exceeding the threshold (let it be 300 sec too), and

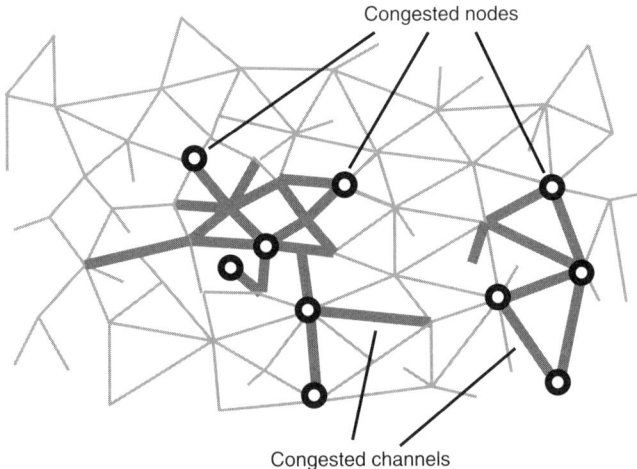

Figure 8.13. Congested links and nodes of the network.

CRISES MANAGEMENT IN OPEN NETWORKS 227

the addresses of such nodes are collected for a separate output (in a direct access to all nodes).

```
direct # start;
USER =
  repeat(
    N == nil; N = 1; Ftime = TIME;
    any # other; Fdiff = TIME - Ftime;
    [Ndiff == nil, Ndiff > Fdiff; Ndiff = Fdiff];
    (Fdiff > 300; BACK_LINK_ADDRESS; ! done), nil
  );
USER = (direct # all; Ndiff > 300; ADDRESS)
```

High-speed and safe bypass channels may be used more efficiently for collecting this information from all nodes.

8.5.5 Inferring Possible Virus Sources

The following solutions are based on the assumption that it is possible to detect, by some special means, from which node the virus has came to the current node, say, using a sophisticated virus check.

Going then to this found predecessor node (which from the beginning could be missed by a general virus check), the procedure is repeated for finding the node from which it was infected itself, and so on, unless no virus predecessor node is detected. This final node, if not the virus source itself, may be regarded as being in some sense closer to the actual virus source than other nodes. There may be more than one such (pseudo-) source in the network, as shown in Figure 8.14 (by stars).

To collect all such nodes as one list in a sorted importance and avoid duplication of them, as the virus network coverage may be a forest rather than a set of chains, we introduce a counter in each infected node accumulating the number of intersections of this node by virus traces starting from nodes with the general alert. Then all nodes with this counter being nonzero (or exceeding some higher threshold) and having no predecessor nodes are collected as a sorted list with aggregate values (as the counter value followed by node address).

```
sequence(
  (direct # all; ? generalCheck;
    repeat(
      Ntrace += 1; Nback = (? detailedCheck);
      Nback != nil; direct # Nback
    )
  ),
  USER = sort(
    direct # all; Nback == nil;
    Ntrace != nil; Ntrace_ADDRESS
  )
)
```

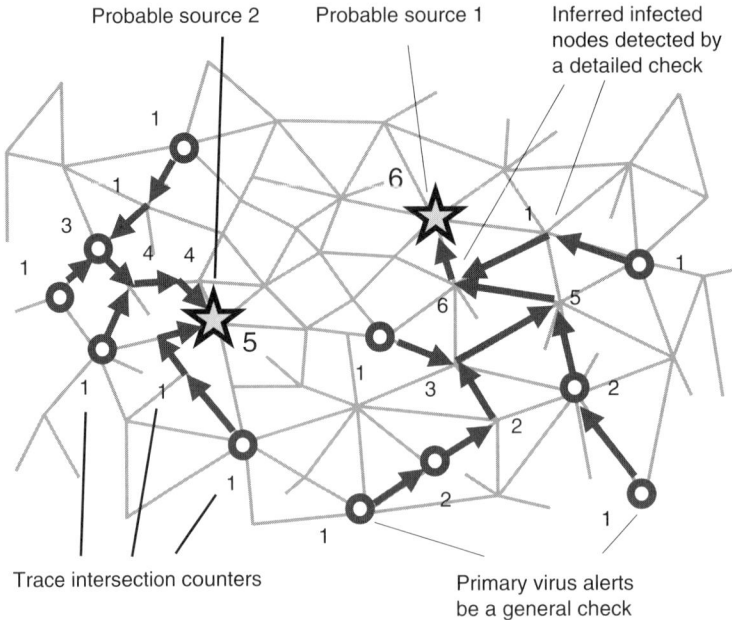

Figure 8.14. Tracing and inferring attack sources.

8.5.6 More Intelligent Solutions Required

Of course, as shown in the previous section, the solution was a very simplified global network analysis procedure, which may not find the real virus source, often thoroughly hidden (just highlighting some language possibilities of distributed network processing). A higher, much more sophisticated and intelligent layer may be needed to be put on top of this spatial virus detection and source inference scenario. This intelligent layer can also be described and implemented in WAVE-WP, with a possible simultaneous access to many other existing, specialized, and advanced virus detection and tracing techniques and systems.

For the advanced network analysis, taking into account that the model can work simultaneously with physical, virtual, and execution worlds, we can first create a detailed virtual representation, or map, of the computer network (which can be kept in a fully distributed form), and then organize its detailed navigation accompanied by corresponding moves in the physical computer network. This may allow us, for example, to find, outline, and investigate completely "dead" regions in the physical network, which could not be seen by the programs described above.

Representing a universal spatial control model based on coordinated code mobility and dynamic tracking, WAVE-WP can also be used for creating essentially new, universal, and intelligent network protocols. These, along with traditional data delivery, will be able to make local and global automatic decisions on the network management, self-recovery after failures, and run time topology restructuring and optimization.

8.6 USING GLOBAL INFRASTRUCTURES IN WAVE-WP

8.6.1 Hypothetical Infrastructure Sketch

WAVE-WP can be applied in a much broader scale, especially for the creation of intelligent national and international infrastructures of different natures, which may widely use automated up to fully automatic control and advanced (cooperative including) robotics. A sketch of a hypothetical national infrastructure (discussed when the author worked in Japan) integrating a variety of crisis management systems and ground, air, and space control centers (manned as well as unmanned), with parallel spatial solutions of nonlocal problems in them, is shown in Figure 8.15.

Such a united system in WAVE-WP, among many other global tasks, may effectively solve the problems of a distributed air defense, where hostile objects penetrating the country's air space can be simultaneously discovered, chased, analyzed, and destroyed using the whole intelligent infrastructure as a collective artificial brain.

Within the unified command-and-control infrastructures provided by the technology, different types of unmanned ground, air, and space vehicles may be used, for example, as possible mobile sensor, relay, or even air traffic management stations, supplementary to the basic ground stations, especially when the latter get damaged or operate inefficiently in crisis situations.

8.6.2 Air Defense Programming Example

Let us consider an elementary code example of tracing and destroying hostile objects penetrating the country's air space, as an extension of the chase of physical objects by mobile intelligence propagating via computer networks, described in the previous chapter.

The following program, discovering all alien objects and assigning individual wave code to follow each of them, regularly informs the external elimination

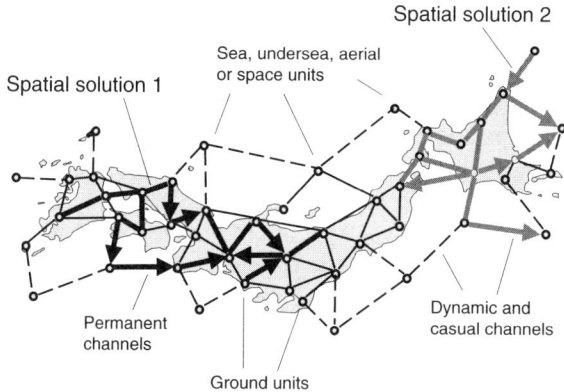

Figure 8.15. Advanced national infrastructure with spatial solutions.

system, let it be named `decideDestroy`, about the identity of the objects seen and also their detection and current time and position. The elimination system chooses among the proposed targets the ones to destroy, which may depend on the danger they currently pose (on the basis of their parameters received), and also current availability of the elimination hardware in the corresponding regions.

```
direct # all; ? alienObjects;
Fobject = VALUE; FstartTime = TIME;
Fidentity = Fobject : 1; FstartCoord = Fobject : 2;
repeat(
  repeat(
    Fidentity ? seen; FcurrentCoord = VALUE;
    free(
      Fidentity_FstartCoord_FcurrentCoord_FstartTime_TIME ?
        decideDestroy
    );
    TIME += 20
  );
  any # any; Fidentity ? seen
)
```

Let us consider some details on the work of this program. The external procedure `alienObjects` leaves a vector of records about the objects seen in the respected nodes of the infrastructure network. The subsequent part of the program self-replicates for each such object, putting the object's identity into variable `Fidentity` and its detection coordinates into variable `FstartCoord`.

At each infrastructure node, while the object is seen by its radars (using external procedure `seen`), its current coordinates are regularly measured, and together with the other parameters (as an aggregate value) are sent to the elimination system under the rule `free`, that is, without waiting for a response from `decideDestroy` and immediately continuing the object's chase by the main program.

If the object is lost in the node, the internal loop terminates, and within the external rule `repeat` a parallel hop to all neighboring infrastructure nodes is organized, in an attempt to see the object from them, after which the internal rule `repeat` starts working again if the object is rediscovered.

The external procedure `decideDestroy` may itself be an arbitrary complex and distributed system, implemented in WAVE-WP or any other technology and being linked to the network of available "killer" vehicles.

Many other important tasks can be efficiently solved using such global infrastructures in WAVE-WP. These may include nationwide or even worldwide simultaneous tracking of movement of funds, goods, transport units, and people (such as, say, illegal drugs, stolen cars, hijacked planes, spying objects, or criminals). Such infrastructures can also be used for inferring, outlining, and eliminating malicious (say, terrorist) infrastructures, global weather forecast, and coordination of complex relief operations after natural or manmade disasters, and so on.

8.7 CONCLUSIONS

We have considered numerous scenarios and problems that may take place and arise in distributed dynamic systems of different natures—from collectively behaving robots—to hosts of Internet cooperatively fighting viruses—to global security and defense infrastructures integrating many other systems and tools, including multiple robots, computer networks, networked radar stations, crisis management centers, and so on.

The WAVE-WP model allows us to comprehend all these systems and scenarios in and over them on a higher, semantic level and organize their effective management with highest integrity and clarity, omitting numerous details that usually obscure the system vision and control. These details can be delegated to be handled by the distributed interpretation system, which can do the job automatically and with high efficiency.

The WAVE-WP programs can start from any system component, dynamically covering and grasping the needed system software and hardware to make it work properly and fulfill the mission objectives, being able to engage new system components or substitute the failed ones at run time, which is of particularly importance for crisis management situations.

9

CONCLUSIONS

A new computational and control model of parallel and distributed nature has been developed. It comprises self-evolving, space-conquering automaton, high-level system navigation and coordination language describing system problems in a spatial pattern-matching mode, and related distributed control mechanisms for management of physical, virtual, and combined worlds. The model allows us to obtain complex spatial solutions in a compact, integral, and seamless way. It can be effectively used for the creation, simulation, processing, management, and control of a variety of dynamic and open systems—from physical to biological, and from artificial to natural.

We summarize here the key issues of the WAVE-WP paradigm and also a number of important application areas, some of them inherited from its predecessor WAVE (Sapaty, 1999a).

9.1 SUMMARY OF THE MAIN FEATURES OF WAVE-WP

9.1.1 Starting from the Whole

The existing distributed computing and control models and technologies follow the analytical approach, representing systems as consisting from pieces, or agents,

Ruling Distributed Dynamic Worlds, by Peter S. Sapaty
ISBN 0-471-65575-9   Copyright © 2005 John Wiley & Sons, Inc.

exchanging messages or sharing distributed objects, with known difficulties of obtaining global behavior from local activities. Overhead in making large distributed systems operate as a single, controllable, entity may be enormous, and the necessity of run time recovery after failures may complicate the situation further. The WAVE-WP model attacks the problem by starting from the opposite side—*from the whole*, allowing us to express it directly, while abstracting from possible system's parts and their interactions, delegating these to an efficient automatic implementation.

9.1.2 The WAVE-WP Automaton

The WAVE-WP automaton may start from any point of the distributed system to be controlled, dynamically covering its parts or the whole, bringing the needed external operations and control to proper locations, and mounting of a variety of parallel and distributed knowledge and control infrastructures. Implanting distributed "soul" into the system organization, the automaton drastically increases the system's integrity, capability of pursuing local and global goals, assessing distributed situations, making autonomous decisions, and recovering from indiscriminate damages. Many spatially cooperating or competing parallel WAVE-WP automata may evolve on the same system's body, serving, say, as deliberative, reactive, and/or reflective spatial processes.

9.1.3 High-Level WAVE-WP Language

The system language expressing full details of this new control automaton has been developed. It has a recursive space-navigating, space-penetrating, and world-conquering nature and can operate with both information and physical matter. The language can also be used as a traditional one, so no integration with (and interfaces to) other programming models and systems may be needed for solving complex distributed knowledge-processing and control problems. Very compact (half-page) syntax of the language makes it particularly suitable for direct interpretation in distributed environments, being supported by effective program code mobility in computer networks. Numerous examples of expressing complex parallel and distributed problems in WAVE-WP have been successfully programmed. Basics of the programming methodology in this spatial pattern-matching mode have been developed.

9.1.4 Distributed WAVE-WP Interpreter

On the implementation layer, the automaton widely uses high-level mobile cooperative program code self-spreading and replicating in networks and can be easily implemented on any existing software or hardware platform. A network of the language interpreters should be embedded into a distributed world to be controlled, installing the interpreters in the most sensitive points. The WAVE-WP interpreter consists of a number of specialized processors working asynchronously and in parallel, handling and sharing specific data structures such as waves queue, incoming and outgoing queues, part of the distributed knowledge network, track forest, and the like and being responsible for different interpretation operations (parser, data processor, control processor, communication processor, sensors, and motion, etc.).

The interpreter can be easily implemented on any existing platform, in software or directly in silicon. The existing public domain WAVE system (in C under Unix/Solaris/Linux) had been used in different countries, especially for distributed network management and simulation of battlefields, with the use of the Internet. The interpreter may also have a physical body, say, as a mobile or humanoid robot, if engaged in operations in the physical world, or can be body-mounted on humans. The whole network of the interpreters can be mobile, changing structure, as robots or humans can be moving at run time.

9.2 SOME MAIN APPLICATION AREAS

9.2.1 Directly Accessing Physical World

Expressing direct movement and operations in the physical world (PW) in WAVE-WP has been tested. The model can describe complex operations in PW on a semantic level, abstracting from implementation, with parallel control directly evolving in physical space, covering the latter. Multiple operations can be performed cooperatively by groups of robots or humans, with transfers of information and physical matter between the interpreters embedded into physical bodies, communicating both electronically and mechanically. Coordinated delivery of physical matter to remote points and its synchronized processing by a group of robots has been programmed. A detailed description of coordinated robotic column movement and its synchronized operation has been shown too.

9.2.2 Distributed Knowledge Processing

Dynamically creating arbitrary knowledge networks in distributed space, which can be modified at run time, WAVE-WP, similar to WAVE, can implement any knowledge processing and control systems in highly parallel and fully distributed way. A program package had been developed in WAVE for basic problems of the graph and network theory, where each graph node can reside on a separate computer (Sapaty, 1999a). The package included finding spanning trees, shortest paths, articulation points, maximum cliques, diameter and radius, and the like, also self-recovering network topologies after indiscriminate damages of their nodes and links. Parallel simulation of other control models in WAVE, like Petri nets, has been demonstrated too.

9.2.3 Operating in Physical World Under the Guidance of Virtual World

Operating in the unity of physical and virtual worlds, the WAVE-WP model can effectively investigate physical worlds and create their reflection in the form of distributed virtual worlds (VW). The latter can guide further movement and search operations in the distributed PW, modifying and updating itself, and so on. Such physical–virtual world unity in one model, where both worlds can be open and dynamic, allows us to create integral intelligent manned or unmanned systems operating effectively in unpredictable environments. This has been demonstrated on an example of optimized parallel territory search by a group of robots sharing distributed knowledge representation of the area of interest, which can be modified and updated (say, reduced or extended) at run time. Parallel operations in PW can be optimized in advance by solving the problems in VW first, in a parallel simulation mode, mapping subsequently or simultaneously the obtained solutions into the PW.

9.2.4 Intelligent Network Management

Inheriting this from WAVE (Sapaty, 1999a), WAVE-WP integrates traditional network management tools and systems, dynamically extracting higher level knowledge from raw data with these tools, and establishing a higher, intelligent, layer allowing us to analyze varying network topologies, regulate network load, and redirect traffic in case of line failures or congestions. Representing a universal spatial control model based on coordinated code mobility and dynamic tracking, WAVE-WP can also be used for creating essentially new, universal, and intelligent network protocols. These, along with traditional data delivery, will be able to make local and global automatic decisions on the network management, self-recovery after failures, and run time topology restructuring and optimization.

9.2.5 Advanced Crisis Reaction Forces

Smaller, dynamic armies, with dramatically increased mobility and lethality, represent nowadays the main direction, also challenge, in the development of advanced military forces oriented on crisis situations. These forces should be capable of conducting nontraditional combat, peacekeeping and recovery operations, withstanding asymmetric threats, and operating in unpredictable environments. They may also effectively use multiple unmanned units, such as mobile robots.

The WAVE-WP technology can quickly assemble a highly operational battle force from dissimilar (and possibly casual) units, setting intelligent command-and-control (CC) infrastructures over them. These infrastructures can follow global mission scenarios and make autonomous decisions in manned, automated, or fully unmanned mode. In case of damages, the technology can either restore the previous CC infrastructure (with, possibly, reduced set of units) or make complete run time reassembling of the entire force, with a new CC infrastructure, taking into account remaining operational units and their physical locations. Both local and global restructuring and reassembling can be carried out at run time, without the loss of overall operational capability.

9.2.6 Massive Cooperative Robotics

Autonomous robotic armies for civil and military applications, comprising thousands of cooperating mobile units, may well become today's reality in WAVE-WP. High-level mission scenarios can be injected into such organizations from any robotic unit, dynamically covering and grasping the whole system. Massive coordinated operations in physical spaces can automatically emerge during parallel interpretation of the WAVE-WP scenarios, while preserving the overall system integrity and ability of pursuing global goals. The scenarios can be represented in such a way that they should (and will be able to) survive by any means, while individual robots may fail, with system intelligence being the feature of the whole campaign, rather than of individual robots.

9.2.7 Distributed Road and Air Traffic Management

Distributed computer networks working in WAVE-WP, covering the space to be controlled, can be efficiently used for both road and air traffic management, as has already been implemented and tested in WAVE (Sapaty, 1999a). The model provides simultaneous tracking of multiple objects in PW by mobile intelligence spreading in the VW, via computer networks. Assigning a personal active mobile code to each object under control brings high flexibility to the distributed control system, with parallel and cooperative tracking of multiple objects and making nonlocal decisions, along with run time optimization and routing. This may especially be important in crisis situations, where a priori flight schedules are becoming useless and the distributed management infrastructure is damaged. The surface roads can be destroyed too, and traffic routing may need to be fully dynamic in order to reach destinations by individual vehicles in the shortest time. Related distributed management scenarios with mobile intelligence have been demonstrated live in WAVE, the predecessor of WAVE-WP.

9.2.8 Autonomous Distributed Cognitive Systems

Cognitive systems belong to the most advanced class of intelligent systems—the ones aware of what they are doing. While cognitive systems include reactive and deliberative processes, they also incorporate mechanisms for self-reflection and adaptive self-modification. The WAVE-WP paradigm allows for the description of interacting deliberative, reactive, and reflective processes on a semantic level, representing the whole mission rather than individual robots. This provides new degrees of freedom for autonomous robotic teams, where collective behavior of robots emerges as a derivative of parallel and distributed interpretation of WAVE-WP language, in the united physical and virtual world. The mission-to-hardware mapping process is fully distributed, not requiring central resources, and each robot may happen to be involved at any level of the distributed command-and-control process in any moment in time. Failed robots can be automatically substituted at run time without loss of the overall mission integrity.

9.2.9 Distributed Interactive Simulation

Inheriting this application from WAVE and having full control over distributed worlds, including their run time creation and modification, WAVE-WP allows for highly efficient, scalable, distributed simulation of complex dynamic systems, such as battlefields, in open computer networks, using a potentially unlimited number of computers working together. Due to full distribution of the simulated space and entities operating in it, there is no need to broadcast changes in terrain or positions of moving entities to other computers, as usual. Each simulated entity operates in its own (current) part of the simulated world, within the range of its sensors, communicating locally with other such entities. The entities can move freely through the simulated space (and between computers, if needed). Using volatile viruslike spatial algorithms in WAVE-WP, a fully dynamic terrain can be effectively modeled in a distributed space. Its parts like clouds, floods, smog, mountains, landslides, and craters can grow, move, and spread seamlessly between computers.

9.2.10 Global Defense and Security

WAVE-WP can also be used in a much broader scale, especially for the creation of intelligent national and international infrastructures of different natures, widely using automated and fully automatic control and advanced robotics. A global system may effectively solve the problems of distributed air defense, where multiple hostile objects penetrating the country's air space can be simultaneously discovered, chased, analyzed, and destroyed using a computerized networked radar system as a collective artificial brain operating in WAVE-WP. In other nonlocal applications, using worldwide computer networks WAVE-WP may effectively discover and trace criminals and their distributed organizations, penetrate into malicious infrastructures studying and eliminating them, with possible additional involvement of special hardware and troops.

9.3 FINAL REMARKS

WAVE-WP allows for a more rational and universal integration, management, and recovery of large complex systems than other approaches—by establishing a higher level of their vision and coordination, symbolically called *overoperability* versus (and in supplement to) the traditional *interoperability*. Distributed system management and coordination scenarios in WAVE-WP are often orders of magnitude simpler and more compact than usual, due to the high level and spatial nature of the model and language. They often help us to see the systems and solutions in them as a whole, avoiding tedious partitioning into parts (agents) and setting their communication and synchronization. These and other routines are effectively shifted to the efficient automatic implementation by dynamic networks of WAVE-WP interpreters. Traditional software and hardware agents are being requested, created, and make sense only when required at certain moments in time, during the spatial development of self-evolving (agentless themselves) mission scenarios.

9.3.1 After the Final Remarks

Usually what we have in computer science or artificial intelligence is mimicking directly or indirectly the human brain and human communications. WAVE-WP may be considered as an exception in this row, being what people do not have and also far more advanced than what they do have. It is a higher level recursive essence covering the world at run time, bringing intelligence into it as a whole, rather than into its individual components. It grasps the very sense of the problems to be solved in distributed worlds, making the whole system self-organize dynamically to follow this sense. The system parts, their communications and command-and-control infrastructures over them appear at lower organizational layers, as a result of this spatial coverage—simplifying by orders of magnitude the management of large, dynamic, and open systems. WAVE-WP really is a means to rule the world!

9.4 FUTURE PLANS

The key issues of both philosophy and implementation of this paradigm are being patented (with sensitive details not present in this text). The technology is also being prepared to meet this century's demands to rule large dynamic systems, as well as new emerging markets, especially related to networked systems and massive, cooperative robotics. A study of numerous potential applications of WAVE-WP (being the main objective of this book) has shown that this "conquering and ruling" philosophy may have much broader influence than technical only, touching in depth the world organization and our ability to make it better. A new book related to these issues, as a sequel to the current as well as the previous one, is being planned, with a provisional title *Grasping the Entirety*.

APPENDIX: WAVE-WP SUMMARY

A.1 EXTENDED LANGUAGE SYNTAX

The full WAVE-WP language syntax is summarized here, which was gradually revealed throughout the whole text of Chapter 3. As was already explained, words in italics in our syntax description represent syntactic categories, braces show zero or more repetitions of the construct with the delimiter shown at the right, if more than one construct, square brackets identify optional constructs, and vertical bars separate alternatives.

Braces, vertical bars, and square brackets, when also used as the WAVE-WP language symbols, are shown in bold. Moreover, for these and all other language symbols (words and separate characters) another font is used, to improve readability.

wave	→	**{** *advance* **;** **}**
advance	→	**{** *move* **,** **}**
move	→	*constant* **|** *variable* **|**
		{ *move act* **}** **|**
		[*rule* **]** **(** *wave* **)** **|**
		[*wave* **]**
constant	→	*information* **|** *physical-matter* **|** *special-constant* **|**
		aggregate-constant
information	→	[*sign*] '**{** *character* **}**' **|** [*sign*] **{ {** *character* **} }**
physical-matter	→	"**{** *character* **}**"
aggregate-constant	→	**{** *constant* **_** **}**
special-constant	→	*number* **|** *place* **|** *address* **|** *time* **|** *speed* **|** *node-kind* **|**
		doer **|** *infra-link* **|** *state* **|** *auxiliary*
number	→	**[** *sign* **]** *integer* **[.** *integer* **]** **[** e **[** *sign* **]** *integer* **]**

integer	→	{ *digit* }
sign	→	+ | −
place	→	{ *dimension number* }
dimension	→	x | y | z | dx | dy | dz | r
address	→	a { *integer* . }
time	→	t { *integer* . } | dt { *integer* . }
speed	→	s *number* | ds *number*
node-kind	→	p | v | e | pv | vp | pe | ve | pve | vpe | ep
doer	→	{ *alphameric* . }
infra-link	→	i *integer*
state	→	abort | thru | out | done | fail
auxiliary	→	direct | [*sign*] any | all | noback | [*sign*] infinite | start | back | fringe | nil
variable	→	*nodal* | *frontal* | *environmental*
nodal	→	M [*alphameric*] | N [*alphameric*]
frontal	→	F [*alphameric*]
environmental	→	KIND | QUALITY | CONTENT | LINK | ORDER | WHERE | TIME | SPEED | DOER | JOINT | ADDRESS | INFRA | BACK | VALUE | PAINT | RESOURCES | USER | GROUND
act	→	*flow-act* | *fusion-act* [=]
flow-act	→	*hop* | *filter* | *state-generator* | *assignment* | *code-injection*
hop	→	# | ##
filter	→	~ | !~ | == | != | < | <= | > | >=
state-generator	→	!
assignment	→	=
code-injection	→	^ | ?
fusion-act	→	*arithmetic* | *vector-special* | *composition*
arithmetic	→	+ | − | ∗ | / | ∗∗
vector-special	→	& | : | ::
composition	→	| | % | && | _
rule	→	*forward-rule* | *echo-rule*
forward-rule	→	*branching* | *repetition* | *synchronization* | *resource-protection*| *network-creation* | *autonomy*
branching	→	sequence | or | orparallel | and | andparallel | random
repetition	→	repeat
synchronization	→	wait
resource-protection	→	grasp
network-creation	→	create
autonomy	→	release | free | quit
echo-rule	→	state | rake | min | max | sort | sum | product | count | none

A.2 COMPACT SYNTAX DESCRIPTION

In a more compact form, oriented on automatic parsing rather than readability, the whole WAVE-WP syntax may be written as follows:

wave	→	{ *advance* ; }
advance	→	{ *move* , }
move	→	*constant* | *variable* | { *move act* } | [*rule*] (*wave*) | [*wave*]
constant	→	[*sign*] '*string*' | [*sign*] { *string* }|"*string*"| *alphameric* | *number* | { *dimension number* } | a { *integer* . } | t{ *integer* . } | dt { *integer* . } | s *number* | ds *number* | p | v | e | pv | vp | pe | ve | pve | vpe | ep | { *alphameric* . } | i *integer* | abort | thru | done | fail| direct | [*sign*] any | all | noback | [*sign*] infinite | start | back | fringe | nil | { *constant* _ }
variable	→	M [*alphameric*] | N [*alphameric*] | F [*alphameric*] | KIND | QUALITY | CONTENT | LINK | ORDER | WHERE | TIME | SPEED | DOER | JOINT | ADDRESS | INFRA | BACK | VALUE | PAINT | RESOURCES | USER | GROUND
act	→	# | ## | ~ | !~ | == | != | < | <= | > | >= | ^ | = | ? | ! | + | − | * | / | ** | & | : | :: | | | % | && | _
rule	→	sequence | or | orparallel | and | andparallel | random | repeat | wait | grasp | create | release | free | quit | state | rake | min | max | sort | sum | product | count | reached | none
dimension	→	x | y | z | dx | dy | dz | r
number	→	[*sign*] *integer* [. *integer*] [e [*sign*] *integer*] |
string	→	{ *character* }
integer	→	{ *digit* }
alphameric	→	{ *letter-or-digit* }
sign	→	+ | −

A.3 PERMITTED ABBREVIATIONS

Shortcuts and substitutions of longer keywords (those of more than three characters long, with few exceptions), similar to the predecessor language WAVE described in the previous book, are allowed in WAVE-WP. These are as follows:

abort	4
ADDRESS	A
andparallel	ap
any	*empty*

BACK	B
back	bk
CONTENT	C
count	ct
create	cr
direct	@
DOER	D
done	1
fail	0
free	fr
fringe	fg
grasp	gr
GROUND	G
infinite	$
INFRA	I
JOINT	J
KIND	K
LINK	L
nil	*empty*
noback	nb
none	nn
ORDER	O
orparallel	op
out	2
PAINT	P
product	pr
QUALITY	Q
quit	qt
rake	rk
random	rd
reached	rc
release	rl
repeat	rp
RESOURCES	R
sequence	sq
sort	sr
SPEED	S
start	st
state	se
TIME	T
thru	3
USER	U
VALUE	V
wait	wt
WHERE	W

REFERENCES

Abbas, A. (2004). *Grid Computing: A Practical Guide to Technology and Applications* (Programming Series), Charles River Media, Hingham, MA, 406p.

Asama, H., T. Arai, T. Fukuda, and T. Hasegawa (2002). *Distributed Autonomous Robotic System 5*, Springer-Verlag Telos.

Attiya, H. and J. Welch (2004). *Distributed Computing: Fundamentals, Simulations, and Advanced Topics*, 2nd ed., (Wiley Series on Parallel and Distributed Computing), Wiley-Interscience, Hoboken, NJ.

Bigus, J. P., J. Bigus, J. Bigus, and J. Bigus (2001). *Constructing Intelligent Agents Using Java: Professional Developer's Guide*, 2nd ed., Wiley, Hoboken, NJ.

Borcea, C., C. Intanagonwiwat, P. Kang, U. Kremer, and L. Iftode (2004). "Spatial Programming Using Smart Messages: Design and Implementation," *Proc. of the 24th International Conference on Distributed Computing Systems (ICDCS 2004)*, Tokyo University of Technology, Tokyo, Japan.

Borst, P. M. (2002). *An Architecture for Distributed Interpretation of Mobile Programs*, Herbert Utz Verlag Wissenschaft, Muenchen.

Braunl, T. (2003). *Embedded Robotics: Mobile Robot Design and Applications with Embedded Systems*, Springer, Berlin, Heidelberg.

Bruce, G. and R. Dempsey (1996). *Security in Distributed Computing: Did You Lock the Door?* Pearson Education.

Butenko, S., R. Murphey, and P. M. Pardalos (2003). *Cooperative Control: Models, Applications, and Algorithms* (Cooperative Systems, Vol. 1), Kluwer Academic.

Carey, G. F. (1997). *Computational Grids: Generations, Adaptation & Solution Strategies*, John Benjamins, Taylor & Francis, Washington, DC.

Comer, D. E. (2000). *Internetworking with TCP/IP, Vol. 1: Principles, Protocols, and Architecture*, 4th ed., Prentice Hall.

Davidson, G. W., M. A. Seaton, and J. Simpson, Eds. (1989). *Chambers Concise Dictionary*, W & R Chambers Ltd. and Cambridge University Press, Cambridge, UK.

Dongarra, J., I. Foster, G. Fox, K. Kennedy, A. White, L. Torczon, and W. Gropp, Eds. (2002). *The Sourcebook of Parallel Computing*, Morgan Kaufmann, San Francisco, CA.

Dudek, G. and M. Jenkin (2000). *Computational Principles of Mobile Robotics*, Cambridge University Press, Cambridge, UK.

Earnshaw, R. and J. Vince (2002). *Intelligent Agents for Mobile and Virtual Media*, Springer-Verlag Telos, London, GB.

Erwin, S. I. (2002). "Incompatible Battle-Command Systems: There's No Easy Fix," *National Defense Magazine*, September.

Ruling Distributed Dynamic Worlds, by Peter S. Sapaty
ISBN 0-471-65575-9 Copyright © 2005 John Wiley & Sons, Inc.

Foster, I. and C. Kesselman (2003). *The Grid 2: Blueprint for a New Computing Infrastructure*, 2nd ed., Morgan Kaufmann.

Fujimoto, R. M. (1999). *Parallel and Distributed Simulation Systems* (Wiley Series on Parallel and Distributed Computing), Wiley-Interscience, New York.

Gage, D. W. (1993). "How to Communicate with Zillions of Robots," *Proc. SPIE Mobile Robots VIII*, Boston, Sept 9–10.

Gonzalez-Valenzuela, S. and V. C. M Leung (2002). "QoS-Routing for MPLS Networks Employing Mobile Agents," *IEEE Network Magazine*, Special Issue, May–June.

Gonzalez-Valenzuela, S. and S. T. Vuong (2002). "Evaluation of Migration Strategies for Mobile Agents in Network Routing," *Proc. of the 4th International Workshop MATA'02*, Barcelona, Spain, October.

Gonzalez-Valenzuela, S., V. C. M. Leung, and S. T. Vuong (2001). "Multipoint-to-point Routing with QoS Guarantees Using Mobile Agents," *Proc. of the 3rd International Workshop MATA'01*, Montreal, Quebec, Canada, August.

Gupta, A., A. Grama, G. Karypis, and V. Kumar (2003). *An Introduction to Parallel Computing: Design and Analysis of Algorithms*, 2nd ed., Addison Wesley, Reading, MA.

Herken, R. (1995). *The Universal Turing Machine: A Half-Century Survey*, 2nd ed., Springer-Verlag.

Holland, J. M. (2003). *Designing Autonomous Mobile Robots: Inside the Mind of an Intelligent Machine*, Newnes.

Hughes, C., T. Hughes, and T. Hughes (2003). *Parallel and Distributed Programming Using C++*, Addison-Wesley, Reading, MA.

Iftode, L., C. Borcea, A. Kochut, C. Intanagonwiwat, and U. Kremer (2003). "Programming Computers Embedded in the Physical World," *Proc. of the 9th IEEE International Workshop on Future Trends of Distributed Computing Systems (FTDCS 2003)*, San Juan, Puerto Rico.

Joseph, J. and C. Fellenstein (2003). *Grid Computing* (On Demand Series), Prentice Hall, Upper Saddle River, NJ.

Lange, D. B. and M. Oshima (1998). *Programming and Deploying Java(TM) Mobile Agents with Aglets(TM)*, Addison-Wesley, Reading, MA.

Langton, C. G., Ed. (1997). *Artificial Life: An Overview (Complex Adaptive Systems)*, reprint edition, MIT Press, Cambridge, MA.

Laudon, K. C. and Jane, P. Laudon (2003). *Management Information Systems*, 8th ed., Prentice Hall, Upper Saddle River, NJ.

Milojicic, D., F. Douglis, and R. Wheeler (1999). *Mobility: Processes, Computers, and Agents* (ACM Press), Addison-Wesley, Reading, MA.

Parker, L. E., G. Bekey, and J. Barhen (2000). *Distributed Autonomous Robotic Systems*, Vol. 4, Springer-Verlag Telos.

Sapaty, P. S. (2004). "The World Processing and Control Technology," *Mathematical Machines and Systems*, No. 3, pp. 3–10.

Sapaty, P. S. (2002). "Over-Operability in Distributed Simulation and Control," *The MSIAC's M&S Journal Online*, Winter Issue, Vol. 4, No. 2, Alexandria, VA, http://www.msiac.dmso.mil/journal/WI03/sap42_1.html.

Sapaty, P. S. (2000a). "Cooperative Exploration of Distributed Worlds in WAVE," *International Journal of Artificial Life and Robotics*, Vol. 4, Springer-Verlag, Tokyo, pp. 109–118.

Sapaty, P. S. (2000b). "Basic Distributed Control Model and Technology for Mobile Crisis Reaction Forces and Their United Air Defense," *Proc. NATO Symp. on System Concepts for Integrated Air Defense of Multinational Mobile Crisis Reaction Forces*, Valencia, Spain, May 22–24.

Sapaty, P. S. (2000c). "High-Level Spatial Scenarios in WAVE," *Proc. of the International Symposium AROB 5th*, Oita, Japan, January, pp. 301–304.

Sapaty, P. S. (1999a). *Mobile Processing in Distributed and Open Environments*, Wiley, New York.

Sapaty, P. S. (1999b). "Cooperative Conquest of Distributed Worlds in WAVE," *Proc. of the Symposium and Exhibition of Unmanned Systems of the New Millennium, AUVSI'99*, Baltimore, MD, July 13–15.

Sapaty, P. S. (1998). "Mobile Programming in WAVE," *Mathematical Machines and Systems*, No. 1, January–March, pp. 3–31.

Sapaty, P. and M. Sugisaka (2001a). "Distributed Artificial Brain for Collectively Behaving Mobile Robots," *Proc. Symposium & Exhibition Unmanned Systems 2001*, July 31–Aug. 2, Baltimore, MD.

Sapaty, P. and M. Sugisaka (2001b). "Towards the Distributed Brain for Collectively Behaving Robots," *Proc. International Conference on Control, Automation and Systems, ICCAS 2001*, October 17–21; Cheju National University, Jeju Island, Korea, pp. 571–574.

Sapaty, P. S. and M. Sugisaka (2002a). "A Language for Programming Distributed Multi-Robot Systems," *Proc. of the Seventh International Symposium on Artificial Life and Robotics (AROB 7th '02)*, January 16–18; B-Com Plaza, Beppu, Oita, Japan, pp. 586–589.

Sapaty, P. and M. Sugisaka (2002b). "Universal Distributed Brain for Mobile Multi-robot Systems," in *Distributed Autonomous Robotic Systems*, H. Asama, T. Arai, T. Fukuda, and T. Hasegava, Eds., Springer, Tokyo, pp. 434–443.

Sapaty, P. and M. Sugisaka (2003a). "Optimized Space Search by Distributed Robotic Teams," *Proc. International Symposium on Artificial Life and Robotics (AROB 8th)*, January 24–26; Beppu, Japan, pp. 189–193.

Sapaty, P. and M. Sugisaka (2003b). "Optimized Space Search by Distributed Robotic Teams," *Proc. World Symposium Unmanned Systems 2003*, July 15–17, Baltimore, MD.

Sapaty, P. and M. Sugisaka (2003c). "Towards a Universal Distributed Brain for Mobile Multi-Robot Systems," *Mathematical Machines and Systems*, No. 3–4, pp. 3–20.

Sapaty, P. and M. Sugisaka (2004). "WAVE-WP (World Processing) Technology," *Proc. First International Conference on Informatics in Control, Automation and Robotics*, Setubal, Portugal, August 25–28, Vol. 1, pp. 92–102.

Sapaty, P., K. Kawamura, M. Sugisaka, and R. Finkelstein (2004a). "Towards Fully Distributed Cognitive Systems," *Proc. Ninth International Symposium on Artificial Life and Robotics (AROB 9th)*, Beppu, Japan, January, pp. 274–278.

Sapaty, P., K. Kawamura, M. Sugisaka, and R. Finkelstein (2004b). "Towards Fully Distributed Cognitive Systems," *Mathematical Machines and Systems*, No. 1, pp. 69–75.

Sapaty, P., V. Klimenko, and M. Sugisaka (2004c). "Dynamic Air Traffic Management Using Distributed Brain Concept," *Proc. Ninth International Symposium on Artificial Life and Robotics (AROB 9th)*, Beppu, Japan, January, pp. 156–159.

Sapaty, P., V. Klimenko, and M. Sugisaka (2004d). "Dynamic Air Traffic Management Using Distributed Brain Concept," *Mathematical Machines and Systems*, No. 1, pp. 3–8.

Sapaty, P., N. Mirenkov, M. Sugisaka, and M. Osano (2004e). "Distributed Artificial Life Using World Processing Technology," *Proc. of the Fifth International Conference on Human and Computer (HC-2004)*, September 1–3; University of Aizu, Japan, pp. 24–29.

Siegwart, R. and I. R. Nourbakhsh (2004). *Introduction to Autonomous Mobile Robots (Intelligent Robotics and Autonomous Agents)*, Bradford Books.

Tanenbaum, A. S. (2002). *Computer Networks*, 4th ed., Prentice Hall, Upper Saddle River, NJ.

Tanenbaum, A. S. and M. van Steen (2002). *Distributed Systems: Principles and Paradigms*, Prentice Hall, Upper Saddle Brook, NJ.

Turing, A. M. (1936). "On Computable Numbers, with an Application to the Entscheidungsproblem," *Proc. London Math. Soc.* Ser. 2, Vol. 42.

Ward, M. (2000). *Virtual Organisms: The Startling World of Artificial Life*, Thomas Dunne Books, St. Martin's Press, New York.

Wheelen, T., J. D. Hunger, and D. Hunger (2003). *Strategic Management and Business Policy*, 9th ed., Prentice Hall, Upper Saddle River, NJ.

Wood, J. (2002). *Java Programming for Spatial Sciences*, CRC Press, Boca Raton, FL.

Wolfram, S. (1994). *Cellular Automata and Complexity*, Perseus Books Group, Westview Press, Boulder, CO.

Wooldridge, M. (2002). *Introduction to MultiAgent Systems*, Wiley, Hoboken, NJ.

INDEX

Abbreviations, 244
`abort`, 65
Access to nodes
  and links, 39
  in the united world, 43
Actions in nodes, 49
Act(s), 60, 73
  flow, 74
  fusion, 81
Addition, 82
`ADDRESS`, 68, 71
Address(ses), 63
  absolute, 47
  mapping, 47
Advance, 60
Aerial object, 202
Agents
  intelligent, 28
  mobile, 28
Aggregation, 84
Air
  defense, programming
    example, 229
  traffic management, 201, 237
Alien networks, 41
`all`, 65
`and`, 87
`andparallel`, 88
`any`, 65
Append, 83
Application
  areas, main, 235–238
  scenario, 26
Arithmetic, 81
  acts, 82
    examples, 82
Artificial life, distributed, 5
Assignment(s), 74, 80
  examples of, 80
Autonomy, 86

`BACK`, 68, 72
`back`, 65
Belongs, 78
Book
  organization, 30
  previous, 26
Branches, splitting into, 86
Branching, 86
  rules
    examples, 88
    semantics, 87
Broadcasting
  to all nodes, 135
  tunnel and surface, 40

Central unit, finding, 192
Centralized, 188
Chain, 188
Cognitive system(s)
  distributed, 214
    autonomous, 237
  dynamic, 213
Code injection, 74, 81
  examples of, 81
Column movement, 169
Command and control (CC)
  basic scenario, 195
  components, intelligent, 3
  implementation, adding payload to, 197
  implementing in WAVE-WP, 196
  recursive hierarchical, 195
Composition, 81
  -decomposition, 83
    examples, 84
  independent or parallel, 61
  parallel-sequential, 61
  sequential, 61
Computation, shifting from, 1
Computer networking, 28
Concatenation, 84

Conclusions, 56, 97, 121, 156, 182, 206, 231
Congested links and nodes, finding, 226
Constant(s), 60
　aggregate, 63
　auxiliary, 65
　examples of, 65
　general on, 62
　special, 62
CONTENT, 68
Contenting, 83
Control
　centralized, using of, 146
　global, setting up, 204
　migrating, 127
Cooperation, between robots, 210
Cooperative robotics, 29
　massive, 236
Coordination of large systems, 1
count, 94
Coverage with rules, 50
create, 91
Crisis (crises)
　management
　　in open networks, 223
　　more scenarios and systems, 207
　　reaction forces, advanced, 236

Data
　definitions, 62
　remote, 53
Defense, global, 238
Degree, 82
Destination regions, 36
direct, 65
Distributed
　algorithms, 6
　artificial life, 5
　cognition, 213
　cognitive systems, 213
　computing, 27
　implementation, 15
　inference, 137
　interpretation issues, 118
　knowledge
　　processing, 235
　　representation and, 21
　management
　　network creation, template-based, 130
　　of road and air traffic, 237
　　solving problems of, 198
　　using dynamic infrastructures, 185
　processing, 21

summation
　cyclic solution, 127
　using migrating control, 127
systems, 27
virtual world, creating, 172
WAVE-WP system, embedding, 223
Distribution
　between particular doers, 131
　of information between doers, 126
　of physical-virtual world, 46
　problems and details, 6
Division, 82
Dynamic patrol, 211
DOER, 68, 71
Doer(s), 45, 64
　and their connections, 45
　and their networks, 99
　-bound, 68
　distribution between, 46
　local operation in, 109
　multiple, 109
　requesting, 114
Does not belong, 78
done, 65
Dynamic
　environments, 99, 201
　hierarchy, 192
　infrastructures, 185
　patrol, 211

Entities
　communicating and impacting, 133
　live, 132
　mobile, 132
　seeing other, 132
Equal, 78
Execution world, 44
Expressions, 95
　examples of, 95

fail, 65
Filter(s), 74, 78
　examples of, 79
Flow-act, 74
free, 92
fringe, 65
Fusion-act, 74
Future combat systems, 220
grasp, 90
Greater, 78
Greater-or-equal, 78

Grid computing, 29
Group movement, modifications, 163
GROUND, 68, 73

Hierarchy, dynamic, 192
Hop(s), 74
  examples of, 77
  simultaneous, 77
  splitting, 77
Hospital
  multirobot service, 216
  robotized, 215
  world representation, 216

Indexing, 83
Information, 62
INFRA, 68, 72
Infrastructure(s)
  creating, from the center, 193
  distributed creation and
    reconfiguration, 186
  dynamic, 185
  global, using in
    WAVE-WP, 229
  hierarchical, 186
  link, 65
    using, 76
  modification, 190
  persistent, creation of and
    moving with, 159
  sketch, 229
Integration
  details, 42
  with frontal variables, 108
  with other systems, 105
Intelligent
  agents, 28
  components, 3
  network management, 236
  solutions, 228
  systems, 3
Interoperability, 2
Interpretation
  distributed, issues, 118
  of expressions, 104
  patterns, 104
Interpreter
  integration with other systems, 105
  main components, 101
Implementation basics, 9
infinite, 65

JOINT, 68, 71

Knowledge
  network(s), 21, 38, 103
    migration between doers, 117
  representation and processing, 21
KIND, 68

Language
  syntax, extended, 241
  top organization, 60
Layer(s), 160
  extending to any number of, 160
Less, 78
Less-or-equal, 78
LINK, 68, 70
Links
  bypass and horizontal, 109
  congested, finding of, 226

Management
  of large systems, 1
  layer, higher, establishing, 224
  scenarios, complex, more, 199
max, 94
Mapping strategies, 118
Merging, 84
min, 94
Mission scenarios, 157
  exemplary, 157
Mobile
  agents, 28
  doers, adding payload, 143
  entities, 132
  robotics, 29
  tracking, 202
Mobility
  in execution world, 139
  in physical world, 140
Move, 60
Movement (moving)
  combined, sequential-parallel, 147
  data through tracks, 116
  in a column, 164, 169
  in EW, 75
  in physical world (PW), 74
    direct, 144
    parallel, 147
    sequential, 145
  into new physical location, 113
  in VW, 75
  of group, 158, 163
  of multiple doers
    by turns, 142
    synchronized, 141

Movement (moving) (*Continued*)
  of single doer, 140
  reverse, or heads-first, 163
  simultaneous, of all nodes, 161
  solutions, different,
        integrating, 166
  stepwise, 158
  to averaged positions, 162
  variants of, 115
Multiplication, 82
Multiple doers, 109
  elementary operations
        involving of, 109
Multirobot
  hospital scenarios, 215
  patrol, as a cognitive system, 214
  service snapshot, 219

Navigation in space, 49
Network(s)
  alien, linking with, 41
  creation, 86, 91
    examples, 92
  management, intelligent, 236
  partitioning, 18
nil, 65
noback, 65
Node(s)
  actions in, 49
  collecting path between, 13
  congested, finding of, 226
  existing, visiting of, 35
  infected, collecting all, 225
  kinds of, 64
  new, 36
    virtual, creating of, 111
  temporary, 34
none, 94
Not equal, 78
Numbers, 63

Observation
  multipoint, 137
  selected, 137
Operations
  elementary, 109
  forward and backward, 106
  local in doers, 109
  spatial, more complex, 115
or, 87
ORDER, 68, 70
orparallel, 87
Other works, 27

out, 65
overhead, 7
Overoperability, 2

PAINT, 68, 72
Pattern with
  arbitrary links, 138
  named and oriented links, 138
Parallel
  and distributed computing, 28
  computing, 27
  distributed machine, 1
Parser, 102
Patrol(ling), 208
  by any number of robots, 211
  by a single robot, 208
  by two robots, 209
Payload, adding
  planting trees, 150
  to CC implementation, 197
Physical
  location, moving into, 113
  matter, 62
    delivery and processing, 55, 167
    transference, 112, 116
    working with, 96
      examples, 96
  neighborhood, 192
  world, 34
    directly accessing, 235
    nodes, temporary, 34
    operating in, 235
    parameters, accessing, 36
    search, 171
Places, 63
Plans, future, 239
Polygons, visiting, 175
Predecessors, recorded, 13
Problem(s) of
  distributed implementation, 15
  distribution, 6
  managing distributed systems, 5
  shortest path, 10
Processor(s)
  communication, 102
  control, 101
  operation, 102
product, 94
Programming
  in a single doer, 123
  in integration of worlds, 151
  in multiple doers, 126
  of virtual world, 130

spatial, 29
    dynamics of, 154
    in WAVE-WP, 123
  traditional, 123

`QUALITY`, 68, 69
Query, elementary, 22, 24
Queue
  incoming, 104
  outgoing, 104
  wave, 102
`quit`, 92, 93

Radar neighborhood
        infrastructure, 201
`rake`, 94
`random`, 88
Region, 35
  patrol, 208, 209
Related areas, 27
`release`, 92, 93
Remarks, final, 238
  after, 239
Remote
  data, returning, 105
  results, 53
  variables, 54
`RESOURCES`, 68, 73
Resource(s)
  management, hierarchical, 198
  protecting common, 90
  protection, 86
      examples, 91
Repetition, 86, 89
  examples of, 89
Ring, 190
Road traffic management, 237
Robotics
  cooperative, massive, 29, 236
  mobile, 29
Robots assigning to
  branches, 177
  scenarios, 169
Route(s), setting specific, 169
Rule(s), 60, 85
  autonomy, 92
      examples, 93
  branching, 86
  coverage with, 50
  echo, 93
      examples, 94
      use of, 149
  forward, 86

  in general, 85
  using, 51

Samples, map-based collection of, 181
Scenario(s)
  cleaning, 218
  fusion-distribution, in WAVE-WP, 222
  life support, 219
  spatial, self-spreading, 10
  state-checking, 217
Search
  combined solutions, 177
  full-depth, 178
  global, for next polygon, 176
  of region, 153
  regular random, 176
  single-step, 175
Security, global, 238
Semantics, direct expression, 17
Sequence, 87
Shortage of vehicles, 118
Shortest path
  collection, 13
      in WAVE-WP, 19
  finding, full program, 20
  problem, 10
  tree finding, 11
      in WAVE-WP, 17
Sets of nodes reached (SNR), 49, 77
Simulation, distributed interactive, 237
Solution(s)
  by two doers, 121
  distributed, 5
  localized, 5
  more intelligent, required, 228
  parallel, 11
  sequential, 120
`sort`, 94
Space
  cleaning scenarios, 174
  modification, run time, 179
  navigation, 49
      parallel solution for, 119
Spanning tree-based collection, 136
Spatial
  automaton, 9
  dynamics, programming of, 154
  operations, more complex, 115
  programming, 29
      in WAVE-WP, 123
`SPEED`, 68, 70
Speed, 44, 64
Splitting, 84

state, 94
State-checking scenario, 217
State(s), 65, 77
  generalization procedure, 85
  generation examples, 80
  generator, 74, 79
  resultant, 50
start, 65
Structuring
  examples of, 61
  with the use of rule, 61
Subtasks, splitting into, 167
Subtraction, 82
sum, 94
  using echo rule, 129
Summation, using central doer, 128
Synchronization, 86, 90,
  adding, 168
  example of, 90
Syntax
  compact, 243
  extended, 241
System(s)
  intelligent, 3
  large distributed, 5
  management, 29
  organization, as a function, 26

Target fusion and distribution, 221
thru, 65
TIME, 68, 70
Time, 44, 64
Topologies, other
  centralized, 187
  chain, 187
  ring, 187
Track(s)
  forest, 103
  infrastructure, 106
  moving data through, 116
  network optimization, 107
  processes, 108
  run time creation, 107
Tracking
  aerial object, 202
  simultaneous multiple, 203
Traffic management
  other tasks, 205
  road and air, 237

United world, 42
  access to nodes, 43
  dynamics, 44
  observation, 153
  planting trees in, 151
USER, 68, 73

VALUE, 68, 72
Value,
  combined, 50
  resultant, 77
Variable(s), 60, 66
  environmental, 68, 103
  frontal, 67, 103
  nodal, 67, 103
Vector(s), 66
  acts special, 83
    examples, 83
  -special, 81
  summation
    parallel, 125
    sequential, 124
Vehicles, dealing with shortage of, 118
Virtual world, 38
  distributed, creating, 172
  guidance, under, 235
Virus sources, inferring, 227

wait, 90
Wave(s), 33, 60
  composition and structuring, 50
  expressions and remote data, 53
  forwarding suspended, 105
  identities, 103
  nature of, 47
  new incoming, 105
  parallel composition, 52
  parallel-sequential composition, 52
  passing through, 105
  queue, 102
  sequential composition, 51
  suspended, 102
WAVE-WP
  advantages of using, 220
  basic ideas, 8
  distributed
    interpretation, 99
    interpreter, 234
  high-level language, 234
  interpreter architecture, 101
  interpreters, universal, 16
  (spatial) automaton, 9, 234
  summary, 241–243
    of the main features, 233–235
WHERE, 68, 70
Whole, 8, 233
  first, 8

World(s)
  and waves, 33
  distribution between doers, 46
  execution, 44
  further integration, 47
  physical, 34
    broadcasting in, 37
    directly operating in, 4
    temporary nodes, 34
  processing language, 59
  virtual, 38
  creating, 130
  distributed
    creating, 172
    inference in, 137
    inhabiting with mobile entities, 132
    mobility in the execution
      world, 139
    observation, 135
    openness, 134
    programming, 130
  united, 42